WEST NOTTS COLLEGE
R0458403

L

TOTAL QUALITY HANDBOOK

DAVID L. GOETSCH

STANLEY B. DAVIS

D1408391

Prentice
Hall

Upper Saddle River, New Jersey
Columbus, Ohio

Library of Congress Cataloging-in-Publication Data
Goetsch, David L.
    Total quality handbook/by David L. Goetsch and Stanley B. Davis.
       p. cm.
    Includes bibliographical references and index.
    ISBN 0-13-027262-0
    1. Total quality management—Handbooks, manuals, etc. 2. Industrial management—Handbooks, manuals, etc. I. Davis, Stanley II. Title.
    HD62.15.G642   2001
    658.4'013—dc21

00-036682
CIP

Vice President and Publisher: Dave Garza
Editor in Chief: Stephen Helba
Executive Editor: Debbie Yarnell
Production Editor: Tricia Huhn
Design Coordinator: Robin G. Chukes
Cover art/photo: ImageBank
Cover Designer: John Jordan
Production Manager: Brian Fox
Marketing Manager: Jimmy Stephens

This book was set in Clearface BT by York Graphic Services Inc. and was printed and bound by R. R. Donnelley & Sons Company. The cover was printed by Phoenix Color Corp.

WEST NOTTINGHAMSHIRE
COLLEGE

658.562

RO 458403

LEARNING RESOURCE
CENTRE

Copyright ©2001 by Prentice-Hall, Inc., Upper Saddle River, New Jersey 07458. All rights reserved. Printed in the United States of America. This publication is protected by Copyright and permission should be obtained from the publisher prior to any prohibited reproduction, storage in a retrieval system, or transmission in any form or by any means, electronic, mechanical, photocopying, recording, or likewise. For information regarding permission(s), write to: Rights and Permissions Department.

Prentice
Hall

10 9 8 7 6 5 4 3 2
ISBN: 0-13-027262-0

Preface

BACKGROUND

The total-quality philosophy is an approach to doing business that focuses all of the resources of the organization on the continual improvement of both quality and competitiveness. To this end, a total-quality organization will continually improve its processes, people, and products. This approach is an effective way and, arguably, the only way to survive and prosper in a globally competitive environment. The global business environment is like the Olympic Games in that at every meeting, the competition is better. Records set at the last Games will be broken at the current Games. Performance that is record-breaking today will be insufficient even to win tomorrow. Consequently, organizations and the people who run them need to understand the total-quality philosophy and how to use it to continually improve everything, every day, forever.

WHY AND FOR WHOM THIS BOOK WAS WRITTEN

This book was written to provide a "handbook" option to complement the authors' book *Quality Management: An Introduction to Total Quality Management for Production, Processing, and Services. Quality Management* is a comprehensive college-level text designed for individuals majoring in quality management or quality engineering at both the undergraduate and graduate levels. *Total Quality Handbook*, on the other hand, was written to be used in the following settings: on-site training programs in business, industry, and government; college classrooms for courses that require a less comprehensive treatment than that provided by *Quality Management* (e.g. quality-related courses that are electives in other majors such as engineering, business, management, or various technology programs); and on-site by managers, supervisors, engineers, and other personnel who are attempting to apply the total-quality philosophy in continually improving their organizations.

ABOUT THE AUTHORS

David L. Goetsch is Provost of the joint campus of the University of West Florida and Okaloosa-Walton Community College in Fort Walton Beach, Florida. He is also president of the Development Institute (TDI), a private company dedicated to the continual improvement of quality, productivity, and competitiveness. Dr. Goetsch is co-founder of The Quality Institute, a partnership of the University of West Florida, Okaloosa-Walton Community College, and the Okaloosa Economic Development Council. He currently serves on the executive board of the Institute.

Stanley B. Davis was a manufacturing executive with Harris Corporation until his retirement in 1991. He was founding managing director of The Quality Institute and is a well-known expert in the areas of implementing total quality, statistical process control, just-in-time manufacturing, benchmarking, ISO 9000, and ISO 14000. He currently serves as professor of quality at the Institute and heads his own firm, Stan Davis Consulting.

Contents

CHAPTER FOURTEEN
Just-In-Time (JIT)   231

Total Quality and Quality Management

The total quality concept as an approach to doing business began to gain wide acceptance in the United States in the late 1980s and early 1990s. However, individual elements of the concept—such as the use of statistical data, teamwork, continual improvement, customer satisfaction, and employee involvement—have been used by visionary organizations for years. It is the pulling together and coordinated use of these and other previously disparate elements that gave birth to the comprehensive concept known as *total quality*. This chapter provides an overview of that concept, laying the foundation for the study of all remaining chapters.

WHAT IS QUALITY?

To understand total quality, one must first understand *quality*. When pressed to define *pornography*, a Supreme Court justice once commented that he couldn't define it but knew it when he saw it. Quality is like that. Although few consumers could define *quality* if asked, all know it when they see it. This makes the critical point that quality is in the eye of the beholder. With the total quality approach, customers ultimately define quality.

People deal with the issue of quality continually in their daily lives. We concern ourselves with quality when grocery shopping, eating in a restaurant, and making a major purchase such as an automobile, a home, a television, or a personal computer. Perceived quality is a major factor by which people make distinctions in the marketplace. Whether we articulate them openly or keep them in the back of our minds, we all apply a number of criteria when making a purchase. The extent to which a purchase meets these criteria determines its quality in our eyes.

One way to understand quality as a consumer-driven concept is to consider the example of eating at a restaurant. How will you judge the quality of the restaurant? Most people apply such criteria as the following:

- Service
- Response time

- Food preparation
- Environment/atmosphere
- Price
- Selection

This example gets at one aspect of quality—the *results* aspect. Does the product or service meet or exceed customer expectations? This is a critical aspect of quality, but it is not the only aspect. *Total quality* is a much broader concept that encompasses not just the results aspect but also the quality of people and the quality of processes.

According to Stephen Uselac, "There is little agreement on what constitutes quality. In its broadest sense, quality is an attribute of a product or service that can be improved. Most people associate quality with a product or service. Quality *is not* only products and services but also includes PROCESSES, ENVIRONMENT, and PEOPLE."[1]

Quality has been defined in a number of different ways by a number of different people and organizations. Consider the following definitions:

- Fred Smith, CEO of Federal Express, defines quality as "performance to the standard expected by the customer."[2]
- The General Services Administration (GSA) defines quality as "meeting the customer's needs the first time and every time."[3]
- Boeing defines quality as "providing our customers with products and services that consistently meet their needs and expectations."[4]
- The U.S. Department of Defense (DOD) defines quality as "doing the right thing right the first time, always striving for improvement, and always satisfying the customer."[5]

In his landmark book *Out of the Crisis*, W. Edwards Deming has this to say about quality:

> Quality can be defined only in terms of the agent. Who is the judge of quality? In the mind of the production worker, he produces quality if he can take pride in his work. Poor quality, to him, means loss of business, and perhaps of his job. Good quality, he thinks, will keep the company in business. Quality to the plant manager means to get the numbers out and to meet specifications. His job is also, whether he knows it or not, continual improvement of leadership.[6]

Deming goes on to make the point that quality has many different criteria and that these criteria change continually.[7] To complicate matters even further, different people value the various criteria differently. For this reason, it is important to measure consumer preferences and to remeasure them frequently. Deming gives an example of the criteria that are important to him in selecting paper:[8]

- It is not slick and, therefore, takes pencil or ink well.
- Writing on the back does not show through.
- It fits into a three-ring notebook.
- It is available at most stationery stores and is, therefore, easily replenished.
- It is reasonably priced.

Each of these preferences represents a variable the manufacturer can measure and use to continually improve decision making. Deming is well known for his belief that 94% of workplace problems are caused by management and especially for his role in helping Japan rise up out of the ashes of World War II to become a major industrial power. Deming's contributions to the quality movement are explained in greater depth later in this chapter.

Although no universally accepted definition of quality exists, enough similarity does exist among the definitions that common elements can be extracted:

- Quality involves meeting or exceeding customer expectations.
- Quality applies to products, services, people, processes, and environments.
- Quality is an ever-changing state (i.e., what is considered quality today may not be good enough to be considered quality tomorrow).

With these common elements extracted, the following definition of *quality* can be set forth:

> *Quality* is a dynamic state associated with products, services, people, processes, and environments that meets or exceeds expectations.

Consider the individual elements of this definition: The *dynamic state* element speaks to the fact that what is considered quality can and often does change as time passes and circumstances are altered. For example, gas mileage is an important criterion in judging the quality of modern automobiles. However, in the days of 20¢-per-gallon gasoline, consumers were more likely to concern themselves with horsepower, cubic inches, and acceleration rates than with gas mileage.

The *products, services, people, processes, and environments* element is critical. It makes the point that quality applies not just to the products and services provided but to the people and processes that provide them and the environments in which they are provided. In the short term, two competitors who focus on continual improvement might produce a product of comparable quality. But the competitor who looks beyond just the quality of the finished product and also focuses on the continual improvement of the people who produce the product, the processes they use, and the environment in which they work will win in the long run and, most frequently, in the short run. This is because quality products are produced most consistently by quality organizations.

THE TOTAL QUALITY APPROACH DEFINED

Just as there are different definitions of *quality*, there are different definitions of *total quality*. For example, the DOD defines the total quality approach as follows:

> TQ consists of continuous improvement activities involving everyone in the organization— managers and workers—in a totally integrated effort toward improving performance at every level. This improved performance is directed toward satisfying such cross-functional goals as quality, cost, schedule, mission, need, and suitability. TQ integrates fundamental management techniques, existing improvement efforts, and technical tools under a disciplined approach focused on continued process improvement. The activities are ultimately focused on increased customer/user satisfaction.[9]

Although this definition holds much value, it would be difficult to repeat should one be asked to define *total quality*. Many of the definitions that have been set forth by various organizations and individuals have this problem. They attempt to incorporate both what total quality is and how it is achieved. By separating those two aspects of total quality, a more workable definition can be achieved.

For the purpose of this book, *total quality* is defined as shown in Figure 1–1. The first part of this definition explains the *what* of total quality; the second part, the *how*. In the case of total quality, the *how* is critical because it is what separates this approach to doing business from others.

The *total* in total quality indicates a concern for quality in the broadest sense— what has come to be known as the "Big Q." Big Q refers to quality of products, services, people, processes, and environments. Correspondingly, "Little Q" refers to a narrower concern that focuses on the quality of one of these elements or individual quality criteria within an individual element.

What It Is
Total quality is an approach to doing business that attempts to maximize the competitiveness of an organization through the continual improvement of the quality of its products, services, people, processes, and environments.

How It Is Achieved
The total quality approach has the following characteristics:
- Strategically based
- Customer focus (internal and external)
- Obsession with quality
- Scientific approach to decision making and problem solving
- Long-term commitment
- Teamwork
- Continual process improvement
- Education and training
- Freedom through control
- Unity of purpose
- Employee involvement and empowerment

Figure 1–1
Total Quality Defined

How Is Total Quality Different?

What distinguishes the total quality approach from traditional ways of doing business can be found in how it is achieved. The distinctive characteristics of total quality are these: customer focus (internal and external), obsession with quality, use of the scientific approach in decision making and problem solving, long-term commitment, teamwork, employee involvement and empowerment, continual process improvement, bottom-up education and training, freedom through control, and unity of purpose, all deliberately aimed at supporting the organizational strategy. Each of these characteristics is explained later in this chapter.

The Historic Development of Total Quality

The total quality movement had its roots in the time and motion studies conducted by Frederick Taylor in the 1920s. Table 1–1 is a timeline that shows some of the major events in the evolution of the total quality movement since the days of Taylor. Taylor is now known as "the father of scientific management."

The most fundamental aspect of scientific management was the separation of planning and execution. Although the division of labor spawned tremendous leaps forward in productivity, it virtually eliminated the old concept of craftsmanship in which one highly skilled individual performed all the tasks required to produce a quality product. In a sense, a craftsman was CEO, production worker, and quality controller all rolled into one person. Taylor's scientific management did away with this by making planning the job of management and production the job of labor. To keep quality from falling through the cracks, it was necessary to create a separate quality department. Such departments had shaky beginnings, and just who was responsible for quality became a clouded issue.

As the volume and complexity of manufacturing increased, quality became an increasingly difficult issue. Volume and complexity together gave birth to quality engineering in the 1920s and reliability engineering in the 1950s. Quality engineering, in turn, resulted in the use of statistical methods in the control of quality, which eventu-

Table 1–1
Selected Historic Milestones in the Quality Movement in the United States

Year	Milestone
1911	Frederick W. Taylor publishes *The Principles of Scientific Management,* giving birth to such techniques as time and motion studies.
1931	Walter A. Shewhart of Bell Laboratories introduces statistical quality control in his book *Economic Control of Quality of Manufactured Products.*
1940	W. Edwards Deming assists the U.S. Bureau of the Census in applying statistical sampling techniques.
1941	W. Edwards Deming joins the U.S. War Department to teach quality control techniques.
1950	W. Edwards Deming addresses Japanese scientists, engineers, and corporate executives on the subject of quality.
1951	Joseph M. Juran publishes the *Quality Control Handbook.*
1961	Martin Company (later Martin-Marietta) builds a Pershing missile that has zero defects.
1970	Philip Crosby introduces the concept of *zero defects.*
1979	Philip Crosby publishes *Quality Is Free.*
1980	Television documentary *If Japan Can . . . Why Can't We?* airs giving W. Edwards Deming renewed recognition in the United States.
1981	Ford Motor Company invites W. Edwards Deming to speak to its top executives, which begins a rocky but productive relationship between the automaker and the quality expert.
1982	W. Edwards Deming publishes *Quality, Productivity, and Competitive Position.*
1984	Philip Crosby publishes *Quality without Tears: The Art of Hassle-Free Management.*
1987	U.S. Congress creates the Malcolm Baldrige National Quality Award.
1988	Secretary of Defense Frank Carlucci directs the U.S. Department of Defense to adopt total quality.
1989	Florida Power and Light wins Japan's coveted Deming Prize, the first non-Japanese company to do so.
1993	The total quality approach is widely taught in U.S. colleges and universities.

ally led to the concepts of *control charts* and *statistical process control,* which are now fundamental aspects of the total quality approach.

Joseph M. Juran, writing on the subject of quality engineering, says:

> This specialty traces its origin to the application of statistical methods for control of quality in manufacture. Much of the pioneering theoretical work was done in the 1920s by the quality assurance department of the Bell Telephone laboratories. The staff members included Shewhart, Dodge, and Edwards. Much of the pioneering application took place (also in the 1920s) within the Hawthorne Works of the Western Electric Company.[10]

Reliability engineering emerged in the 1950s. It began a trend toward moving quality control away from the traditional after-the-fact approach and toward inserting it throughout design and production. However, for the most part, quality control in the 1950s and 1960s involved inspections that resulted in nothing more than cutting out bad parts.

World War II had an impact on quality that is still being felt. In general, the effect was negative for the United States and positive for Japan. Because of the urgency to meet production schedules during the war, U.S. companies focused more on meeting delivery dates than on quality. This approach became a habit that carried over even after the war.

Japanese companies, on the other hand, were forced to learn to compete with the rest of the world in the production of nonmilitary goods. At first their attempts were unsuccessful, and "Made in Japan" remained synonymous with poor quality, as it had been before World War II. Around 1950, however, Japan decided to get serious about quality and establishing ways to produce quality products. Here is how Juran describes the start of the Japanese turnaround:

> To solve their quality problems the Japanese undertook to learn how other countries managed for quality. To this end the Japanese sent teams abroad to visit foreign companies and study their approach, and they translated selected foreign literature into Japanese. They also invited foreign lecturers to come to Japan and conduct training courses for managers.
>
> From these and other inputs the Japanese devised some unprecedented strategies for creating a revolution in quality. Several of those strategies were decisive:
>
> 1. The upper managers personally took charge of leading the revolution.
> 2. All levels and functions underwent training in managing for quality.
> 3. Quality improvement was undertaken at a continuing, revolutionary pace.
> 4. The workforce was enlisted in quality improvement through the QC-concept.[11]

More than any other single factor. It was the Japanese miracle—which was not a miracle at all but the result of a concerted effort that took 20 years to really bear fruit—that got the rest of the world to focus on quality. When Western companies finally realized that quality was the key factor in global competition, they responded. Unfortunately, their first responses were the opposite of what was needed.

Juran describes those initial responses as follows:

> The responses to the Japanese quality revolution took many directions. Some of these directions consisted of strategies that had no relation to improving American competitiveness in quality. Rather, these were efforts to block imports through restrictive legislation and quotas, criminal prosecutions, civil lawsuits, and appeals to buy American.[12]

In spite of these early negative reactions, Western companies began to realize that the key to competing in the global marketplace was to improve quality. With this realization, the total quality movement finally began to gain momentum.

KEY ELEMENTS OF TOTAL QUALITY

The total quality approach was defined in Figure 1–1. This definition has two components: the *what* and the *how* of total quality. What distinguishes total quality from other approaches to doing business is the *how* component of the definition. This component has eleven critical elements, each of which is explained in the remainder of this section.

Strategically Based

Total quality organizations have a comprehensive strategic plan that contains at least the following elements: vision, mission, broad objectives, and activities that must be completed to accomplish the broad objectives. The strategic plan of a total quality organization is designed to give it a *sustainable competitive advantage* in the marketplace. The competitive advantages of a total quality organization are geared toward achieving world-leading quality and improving on it continually forever.

Customer Focus

In a total quality setting, the customer is the driver. This point applies to both internal and external customers. External customers define the quality of the product or service delivered. Internal customers help define the quality of the people, processes, and environments associated with the products or services.

Quality and teamwork expert Peter R. Scholtes explains the concept of *customer focus* as follows:

Whereas Management by Results begins with profit and loss and return on investment, Quality Leadership starts with the customer. Under Quality Leadership, an organization's goal is to meet and exceed customer needs, to give lasting value to the customer. The return will follow as customers boast of the company's quality and service. Members of a quality organization recognize both external customers, those who purchase or use the products or services, and internal customers, fellow employees whose work depends on the work that precedes them.[13]

Obsession with Quality

In a total quality organization, internal and external customers define quality. With quality defined, the organization must then become obsessed with meeting or exceeding this definition. This means all personnel at all levels approach all aspects of the job from the perspective of "How can we do this better." When an organization is obsessed with quality, good enough is never good enough.

Scientific Approach

Total-quality detractors sometimes view total quality as nothing more than "mushy people stuff."[14] Although it is true that people skills, involvement, and empowerment are important in a total quality setting, they represent only a part of the equation. Another important part is the use of the scientific approach in structuring work and in decision making and problem solving that relates to the work. This means that hard data are used in establishing benchmarks, monitoring performance, and making improvements.

Long-Term Commitment

Organizations that implement management innovations after attending short-term seminars often fail in their initial attempt to adopt the total quality approach. This is because they approach total quality as just another management innovation rather than as a whole new way of doing business that requires a whole new corporate culture.

Too few organizations begin the implementation of total quality with the long-term commitment to change that is necessary for success. Quality consultant Jim Clemmer of Toronto-based Achieve International describes mistakes that organizations frequently make when starting quality initiatives, the first of which is as follows:

> Senior managers decide they want all of the benefits of total quality, so they hire an expert or throw some money at a particular department. Why that approach doesn't work has been widely discussed; I won't belabor the point.[15]

Teamwork

In traditionally managed organizations, the best competitive efforts are often between departments within the organization. Internal competition tends to use energy that should be focused on improving quality and, in turn, external competitiveness. Scholtes describes the need for teamwork as follows:

> Where once there may have been barriers, rivalries, and distrust, the quality company fosters teamwork and partnerships with the workforce and their representatives. This partnership is not a pretense, a new look to an old battle. It is a common struggle for the customers, not separate struggles for power. The nature of a common struggle for quality also applies to relationships with suppliers, regulating agencies, and local communities.[16]

Continual Process Improvement

Products are developed and services delivered by people using processes within environments (systems). To continually improve the quality of products or services—which is a fundamental goal in a total quality setting—it is necessary to continually improve systems.

Education and Training

Education and training are fundamental to total quality because they represent the best way to improve people on a continual basis. According to Scholtes:

> In a quality organization everyone is constantly learning. Management encourages employees to constantly elevate their level of technical skill and professional expertise. People gain an ever-greater mastery of their jobs and learn to broaden their capabilities.[17]

It is through education and training that people who know how to work hard learn how to also work smart.

Freedom through Control

Involving and empowering employees is fundamental to total quality as a way to simultaneously bring more minds to bear on the decision-making process and increase the ownership employees feel in decisions that are made. Total quality detractors sometimes mistakenly see employee involvement as a loss of management control, when in fact control is fundamental to total quality. The freedoms enjoyed in a total quality setting are actually the result of well-planned and carried-out controls.

Scholtes explains this paradox as follows:

> In quality leadership there is control, yet there is freedom. There is control over the best-known method for any given process. Employees standardize processes and find ways to ensure everyone follows the standard procedures. They reduce variation in output by reducing variation in the way work is done. As these changes take hold, they are freer to spend time eliminating problems, to discover new markets, and to gain greater mastery over processes.[18]

Unity of Purpose

Historically, management and labor have had an adversarial relationship in U.S. industry. One could debate the reasons behind management–labor discord ad infinitum without achieving consensus. From the perspective of total quality, who or what is to blame for adversarial management–labor relations is irrelevant. What is important is this: To apply the total quality approach, organizations must have unity of purpose. This means that internal politics has no place in a total quality organization. Rather, collaboration should be the norm.

A question frequently asked concerning this element of total quality is "Does unity of purpose mean that unions will no longer be needed?" The answer is that unity of purpose has nothing to do with whether unions are needed. Collective bargaining is about wages, benefits, and working conditions, not about corporate purpose and vision. Employees should feel more involved and empowered in a total quality setting than in a traditionally managed situation, but the goal of total quality is to enhance competitiveness, not to eliminate unions. For example, in Japan, where companies are known for achieving unity of purpose, unions are still very much in evidence. Unity of purpose does not necessarily mean that labor and management will always agree on wages, benefits, and working conditions.

Employee Involvement and Empowerment

Employee involvement and empowerment is one of the most misunderstood elements of the total quality approach and one of the most misrepresented by its detractors. The basis for involving employees is twofold. First, it increases the likelihood of a good decision, a better plan, or a more effective improvement by bringing more minds to bear on the situation—not just any minds, but the minds of the people who are closest to the work in question. Second, it promotes ownership of decisions by involving the people who will have to implement them.

Empowerment means not just involving people but involving them in ways that give them a real voice. One of the ways this can be done is by structuring work that allows employees to make decisions concerning the improvement of work processes within well-specified parameters. Should a machinist be allowed to unilaterally drop a vendor if the vendor delivers substandard material? No. However, the machinist should have an avenue for offering his or her input into the matter.

Should the same machinist be allowed to change the way she sets up her machine? If by so doing she can improve her part of the process without adversely affecting someone else's, yes. Having done so, her next step should be to show other machinists her innovation so that they might try it.

CONTRIBUTIONS OF DEMING AND JURAN

The two principal American quality pioneers are W. Edwards Deming and Joseph M. Juran. Deming is known primarily for the Deming Cycle, his Fourteen Points, and the Seven Deadly Diseases. Juran is known for his Three Basic Steps to Progress, Ten Steps to Quality Improvement, The Pareto Principle, and The Juran Trilogy.

The Deming Cycle

Summarized in Figure 1–2, the Deming Cycle was developed to link the production of a product with consumer needs and focus the resources of all departments (research, design, production, marketing) in a cooperative effort to meet those needs. The Deming Cycle proceeds as follows:

Figure 1–2
The Deming Cycle

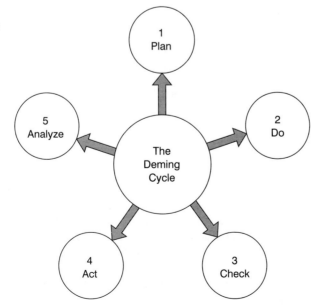

1. Conduct consumer research and use it in planning the product (plan).
2. Produce the product (do).
3. Check the product to make sure it was produced in accordance with the plan (check).
4. Market the product (act).
5. Analyze how the product is received in the marketplace in terms of quality, cost, and other criteria (analyze).

Deming's Fourteen Points

Deming's philosophy is both summarized and operationalized by his Fourteen Points. Scholtes describes them as follows:

> Over the years, Dr. Deming has developed 14 points that describe what is necessary for a business to survive and be competitive today. At first encounter, their meaning may not be clear. But they are the very heart of Dr. Deming's philosophy. They contain the essence of all his teachings. Read them, think about them, talk about them with your co-workers or with experts who deeply understand the concepts. And then come back to think about them again. Soon you will start to understand how they work together and their significance in the true quality organization. Understanding the 14 points can shape a new attitude toward work and the work environment that will foster continuous improvement.[19]

Deming's Fourteen Points are contained in Figure 1–3. Deming has modified the specific wording of various points over the years, which accounts for the minor differences among the Fourteen Points as described in various publications. Deming has stated repeatedly in his later years that if he had it all to do over again, he would leave off the numbers. Many people, in Deming's opinion, interpret numbers as an order of priority or progression when this, in fact, is not the point: the numbers represent neither an order of progression nor relative priorities.[20]

---

1. Create constancy of purpose toward the improvement of products and services in order to become competitive, stay in business, and provide jobs.
2. Adopt the new philosophy. Management must learn that it is a new economic age and awaken to the challenge, learn their responsibilities, and take on leadership for change.
3. Stop depending on inspection to achieve quality. Build in quality from the start.
4. Stop awarding contracts on the basis of low bids.
5. Improve continuously and forever the system of production and service, to improve quality and productivity, and thus constantly reduce costs.
6. Institute training on the job.
7. Institute leadership. The purpose of leadership should be to help people and technology work better.
8. Drive out fear so that everyone may work effectively.
9. Break down barriers between departments so that people can work as a team.
10. Eliminate slogans, exhortations, and targets for the workforce. They create adversarial relationships.
11. Eliminate quotas and management by objectives. Substitute leadership.
12. Remove barriers that rob employees of their pride of workmanship.
13. Institute a vigorous program of education and self-improvement.
14. Make the transformation everyone's job and put everyone to work on it.

Figure 1–3
Deming's Fourteen Points

Deming's Seven Deadly Diseases

The Fourteen Points summarize Deming's views on what a company must do to effect a positive transition from business as usual to world-class quality. The Seven Deadly Diseases summarize the factors that he believes can inhibit such a transformation (see Figure 1–4).

The description of these factors rings particularly true when viewed from the perspective of U.S. firms trying to compete in the global marketplace. Some of these factors can be eliminated by adopting the total quality approach, but three cannot. This does not bode well for U.S. firms trying to regain market share. Total quality can eliminate or reduce the impact of a lack of consistency, personal review systems, job hopping, and using only visible data. However, total quality will not free corporate executives from pressure to produce short-term profits, excessive medical costs, or excessive liability costs. These are diseases of the nation's financial, health care, and legal systems, respectively.

By finding ways for business and government to cooperate appropriately without collaborating inappropriately, other industrialized countries have been able to focus their industry on long-term rather than short-term profits, hold down health care costs, and prevent the proliferation of costly litigation that has occurred in the United States. Excessive health care and legal costs represent non-value-added costs that must be added to the cost of products produced and services delivered in the United States.

Juran's Three Basic Steps to Progress

Juran's Three Basic Steps to Progress (listed in Figure 1–5) are broad steps that, in Juran's opinion, companies must take if they are to achieve world-class quality. He also believes there is a point of diminishing return that applies to quality and competitiveness. An example illustrates his observation:

> Say that an automobile maker's research on its midrange line of cars reveals that buyers drive them an average of 50,000 miles before trading them in. Applying Juran's theory, this automaker should invest the resources necessary to make this line of cars run trouble-free for perhaps 60,000 miles. According to Juran, resources devoted to improving quality beyond this point will run the cost up higher than the typical buyer is willing to pay.

---

1. Lack of constancy of purpose to plan products and services that have a market sufficient to keep the company in business and provide jobs.
2. Emphasis on short-term profits; short-term thinking that is driven by a fear of unfriendly takeover attempts and pressure from bankers and shareholders to produce dividends.
3. Personal review systems for managers and management by objectives without providing methods or resources to accomplish objectives. Performance evaluations, merit ratings, and annual appraisals are all part of this *disease*.
4. Job hopping by managers.
5. Using only visible data and information in decision-making with little or no consideration given to what is not known or cannot be known.
6. Excessive medical costs.
7. Excessive costs of liability driven up by lawyers that work on contingency fees.

Figure 1–4
Deming's Seven Deadly Diseases

1. Achieve structured improvements on a continual basis combined with dedication and a sense of urgency.
2. Establish an extensive training program.
3. Establish commitment and leadership on the part of higher management.

Figure 1–5
Juran's Three Basic Steps to Progress
Stephen Uselac, *Zen Leadership: The Human Side of Total Quality Team Management* (Loudonville, OH: Mohican, 1993), 37.

Juran's Ten Steps to Quality Improvement

Examining Juran's Ten Steps to Quality Improvement (in Figure 1–6), you will see some overlap between them and Deming's Fourteen Points. They also mesh well with the philosophy of quality experts whose contributions are explained later in this chapter.

The Pareto Principle

The Pareto Principle espoused by Juran shows up in the views of most quality experts, although it often goes by other names, such as Juran's 80/20 Rule (in Figure 1–7). According to this principle, organizations should concentrate their energy on eliminating the vital few sources that cause the majority of problems. Further, both Juran and Deming believe that systems that are controlled by management are the systems in which the majority of problems occur.

The Juran Trilogy

The Juran Trilogy (Figure 1–8) summarizes the three primary managerial functions. Juran's views on these functions are explained in the following sections.[21]

Quality Planning

Quality planning involves developing the products, systems, and processes needed to meet or exceed customer expectations. The following steps are required:

1. Build awareness of both the need for improvement and opportunities for improvement.
2. Set goals for improvement.
3. Organize to meet the goals that have been set.
4. Provide training.
5. Implement projects aimed at solving problems.
6. Report progress.
7. Give recognition.
8. Communicate results.
9. Keep score.
10. Maintain momentum by building improvement into the company's regular systems.

Figure 1–6
Juran's Ten Steps to Quality Improvement
Stephen Uselac, *Zen Leadership: The Human Side of Total Quality Team Management* (Loudonville, OH: Mohican, 1993), 37.

The Pareto Principle

"This principle is sometimes called the 80/20 rule: 80% of the trouble comes from 20% of the problems. Though named for turn-of-the-century economist Vilfredo Pareto, it was Dr. Juran who applied the idea to management. Dr. Juran advises us to concentrate on the 'vital few' sources of problems and not be distracted by those of lesser importance."

Source: Peter R. Scholtes, *The Team Handbook* (Madison, WI: Joiner Associates, 1992), 2-9.

Figure 1–7
Juran's 80/20 Rule

1. Determine who the customers are.
2. Identify customers' needs.
3. Develop products with features that respond to customer needs.
4. Develop systems and processes that allow the organization to produce these features.
5. Deploy the plans to operational levels.

Quality Control

The control of quality involves the following processes:

1. Assess actual quality performance.
2. Compare performance with goals.
3. Act on differences between performance and goals.

Quality Improvement

The improvement of quality should be ongoing and continual:

1. Develop the infrastructure necessary to make annual quality improvements.
2. Identify specific areas in need of improvement, and implement improvement projects.
3. Establish a project team with responsibility for completing each improvement project.

Figure 1–8
The Juran Trilogy
The Juran Trilogy® is a registered
trademark of Juran Institute, Inc.

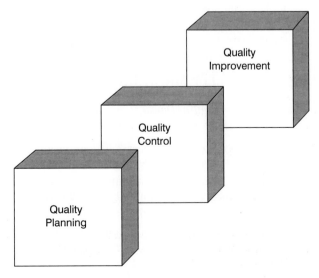

4. Provide teams with what they need to be able to diagnose problems to determine root causes, develop solutions, and establish controls that will maintain gains made.

WHY TOTAL QUALITY EFFORTS SOMETIMES FAIL

Organizations that succeed never approach total quality as just another management innovation or, even worse, as a quick fix. Rather, they approach total quality as a new way of doing business. What follows are common errors organizations make when implementing total quality. The successful organizations avoid these errors.

- *Senior management delegation and poor leadership.*   Some organizations attempt to start a quality initiative by delegating responsibility to a hired expert rather than applying the leadership necessary to get everyone involved.
- *Team mania.*   Ultimately teams should be established, and all employees should be involved with them. However, working in teams is an approach that must be learned. Supervisors must learn how to be effective coaches and employees must learn how to be team players. The organization must undergo a cultural change before teamwork can succeed. Rushing in and putting everyone in teams before learning has occurred and the corporate culture has changed will create problems rather than solve them.
- *Deployment process.*   Some organizations develop quality initiatives without concurrently developing plans for integrating them into all elements of the organization (i.e., operations, budgeting, marketing, etc.). According to Jim Clemmer, "More time must be spent preparing plans and getting key stake holders on board, including managers, unions, suppliers, and other production people. It takes time to pull them in. It involves thinking about structure, recognition, skill development, education, and awareness."[22]
- *Taking a narrow, dogmatic approach.*   Some organizations are determined to take the Deming approach, Juran approach, or Crosby approach and use only the principles prescribed in them. None of the approaches advocated by these and other leading quality experts is truly a one-size-fits-all proposition. Even the experts encourage organizations to tailor quality programs to their individual needs.
- *Confusion about the differences among education, awareness, inspiration, and skill building.*   According to Clemmer, "You can send people to five days of training in group dynamics, inspire them, teach them managerial styles, and show them all sorts of grids and analysis, but that doesn't mean you've built any skills. There is a time to educate and inspire and make people aware, and there is a time to give them practical tools they can use to do something specific and different than they did last week."[23]

THE FUTURE OF QUALITY MANAGEMENT

In an article for *Quality Digest*,[24] Armand Feigenbaum explains several trends that will shape the future of quality management. Those trends are as follows:

- *Demanding global customers.*   The provision of quality begets an ever-increasing demand for quality. Today's customers share two common characteristics: (a) they are part of regional trade alliances such as the Americas, Europe, and Asia; and (b) they expect both high quality and added value.
- *Shifting customer expectations.*   Increasingly, today's global customer is interested not just in the quality of a product provided but also the quality of the organization that backs it up. Customers want an excellent product or service from an organization that also provides accurate billing, reliable delivery, and after-purchase support.

■ *Opposing economic pressures.* The global marketplace exerts enormous, unrelenting pressure on organizations to continually improve quality while simultaneously reducing the prices they charge for goods and services. The key to achieving higher quality and lower prices for customers is the reduction of the expenses associated with satisfying unhappy customers—expenses that amount to as much as 25% of the cost of sales in many companies.

■ *New approaches to management.* Companies that succeed in the global marketplace have learned that *you manage budgets, but lead people.* The old approach of providing an occasional seminar or motivational speech for employees without making any fundamental changes in the way the organization operates will no longer work.

Quality Management Characteristics for the Future

To succeed in the global marketplace for now and in the future, organizations will have to operate according to the principles of quality management. Such companies will have the following characteristics:

■ A total commitment to continually increasing value for customers, investors, and employees

■ A firm understanding that *market driven* means that quality is defined by customers, not the company

■ A commitment to *leading* people with a bias for continuous improvement and communication

■ A recognition that sustained growth requires the simultaneous achievement of four objectives all the time, forever: (a) customer satisfaction, (b) cost leadership, (c) effective human resources, and (d) integration with the supplier base.

■ A commitment to fundamental improvement through knowledge, skills, problem solving, and teamwork.

Companies that develop these characteristics will be those that fully institutionalize the principles of quality management. Consequently, quality management as both a practice and a profession has a bright future. In fact, in terms of succeeding in the global marketplace, quality management is the future. Consequently, more and more companies are making quality management the way they do business, and more and more institutions of higher education are offering quality management courses and programs.

ENDNOTES

1. Stephen Uselac, *Zen Leadership: The Human Side of Total Quality Team Management* (Loudonville, OH: Mohican, 1993), 20.
2. Air Force Development Test Center, *Total Quality Management (TQM) Training Package*, 1991, 8.
3. Jerry Romano, "It's Time for a Quality Management Revolution," workshop presented to the Emerald Coast Personnel Manager's Association, 20 May 1992.
4. Romano, 1.
5. Romano, 1.
6. W. Edwards Deming, *Out of the Crisis* (Cambridge: Massachusetts Institute of Technology Center for Advanced Engineering Study, 1986), 168.
7. Deming, 169.
8. Deming, 169.
9. Air Force Development Test Center, Elgin Air Force Base, Total Quality Management training package, 1991, 13.

10. J. M. Juran, *Juran on Leadership for Quality* (New York: Free Press, 1989), 5.
11. Juran, 7–8.
12. Juran, 10.
13. Peter R. Scholtes, *The Team Handbook* (Madison, WI: Joiner Associates, 1992), 1–11.
14. Comment by a participant in a workshop on total quality presented by David Goetsch in Fort Walton Beach, Florida, in January 1993.
15. Jim Clemmer, "Eye on Quality," *Total Quality Newsletter* (3 April 1992): 7.
16. Scholtes, 1–13.
17. Scholtes, 1–13.
18. Scholtes, 1–12.
19. Scholtes, 2–4.
20. W. Edwards Deming, comments made during a teleconference on total quality broadcast by George Washington University in Spring 1992.
21. Juran, 20.
22. Jim Clemmer, "5 Common Errors Companies Make Starting Quality Initiatives," *Total Quality* 3 (April 1992): 7.
23. Clemmer, 7.
24. Armand V. Feigenbaum, "The Future of Quality Management," *Quality Digest*, 18(4) (May 1998): 33–38.

Strategic Planning

STRATEGIC PLANNING OVERVIEW

Strategic planning is the process whereby organizations develop a vision, mission, guiding principles, broad objectives, and specific strategies for achieving the broad objectives. Before even beginning the planning process, an organization should conduct a SWOT analysis. SWOT is the acronym for *strengths, weaknesses, opportunities,* and *threats.* A SWOT analysis answers the following questions: What are this organization's strengths? What are this organization's weaknesses? What opportunities exist in this organization's business environment? What threats exist in this organization's business environment?

The steps in the strategic planning process (Figure 2–1) should be completed in order, because each successive step grows out of the preceding step. The SWOT analysis provides a body of knowledge that is needed to undertake strategic planning. The mission grows out of and supports the vision. The guiding principles, which represent the organization's value system, guide the organization's behavior as it pursues its mission. The broad objectives grow out of the mission and translate it into measurable terms. Specific tactics tie directly to the broad objectives. Typically there will be two to five strategies for each objective, but this is a rule of thumb, not a hard and fast rule.

CONDUCTING THE SWOT ANALYSIS

The rationale for conducting a SWOT analysis before proceeding with the development of the strategic plan is that the organization's plan should produce a good fit between its internal situation and its external situation. An organization's internal situation is defined by its strengths and weaknesses. An organization's external situation is defined by the opportunities and threats that exist in its business environment. The strategic plan should be designed in such a way that it exploits an organization's strengths and

Figure 2–1
The Strategic Planning Process

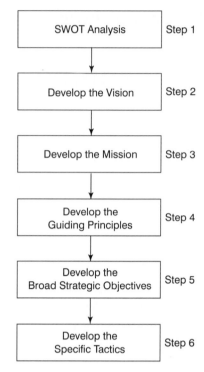

SWOT Analysis	Step 1
Develop the Vision	Step 2
Develop the Mission	Step 3
Develop the Guiding Principles	Step 4
Develop the Broad Strategic Objectives	Step 5
Develop the Specific Tactics	Step 6

opportunities, while simultaneously overcoming, accommodating, or circumventing weaknesses and threats.

Identifying Organizational Strengths

An organizational strength is any characteristic or capability that gives the organization a competitive advantage. The following are examples of common organizational strengths:

- Financial strength
- A good reputation in the marketplace
- Strategic focus
- High-quality products/services
- Proprietary products/services
- Cost leadership
- Strong management team
- Efficient technological processes
- Talented workforce
- Faster time to market

These are just some of the strengths an individual organization may have; many others are possible. The key is accurately defining an organization's strengths before beginning to develop its strategic plan.

Identifying Organizational Weaknesses

An organizational weakness is any characteristic or capability that is lacking to the extent that it puts the organization at a competitive disadvantage. These are examples of common organizational weaknesses:

- Strategic confusion/lack of direction
- Obsolete facilities
- Obsolete processes
- Weak management team
- Insufficient skills/capabilities in the workforce
- Poorly defined operating procedures
- Too narrow a product line
- Products with decreasing demand
- Too diverse a product line
- Poor image in the marketplace
- Weak distribution system
- Weak financial position
- High unit costs compared with those of competitors
- Poor quality in products/services

These are just a few of many weaknesses an organization may have. The main thing is to identify an organization's weaknesses accurately before undertaking the strategic planning process.

Identifying External Opportunities

External opportunities are opportunities in the organization's business environment that represent potential avenues for growth and/or gaining a sustainable competitive advantage. The following are examples of external opportunities that organizations may have:

- Availability of new customers
- An expanding market for existing or potential/planned products
- Ability to diversify into related products/services
- Removal of growth-inhibitive barriers
- Failures of competitors
- New productivity or quality-enhancing technologies coming on line.

Of course, other external opportunities might be available to an organization besides these. You need to identify all such opportunities accurately before undertaking the strategic planning process.

Identifying External Threats

An external threat is a phenomenon in an organization's business environment that has the potential to put the organization at a competitive disadvantage. Such external threats might include the following:

- Entry of lower-cost competitors
- Entry of higher-quality competitors
- Increased sales of substitute products/services
- Significant slowdown in market growth
- Introduction of costly new regulatory requirements
- Poor supplier relations
- Changing tastes and habits of consumers
- Potentially damaging demographic changes

Many other external threats might confront an organization. Accurately identifying every potential external threat before you begin the strategic planning process is a must.

DEVELOPING THE VISION

An organization's guiding force, the dream of what it wants to become, and its reason for being should be apparent in its vision. A vision is like a beacon in the distance toward which the organization is always moving. Everything about the organization—its structure, policies, procedures, and allocation of resources—should support the realization of the vision.

In an organization with a clear vision, it is relatively easy to stay appropriately focused. If a policy does not support the vision, why have it? If a procedure does not support the vision, why adopt it? If an expenditure does not support the vision, why make it? If a position or even department doesn't support the vision, why keep it? An organization's vision must be established and articulated by executive management and understood by all employees. The first step in articulating an organizational vision is writing it down. This is called the *vision statement.*

Writing the Vision Statement

A well-written vision statement, regardless of the type of organization, has the following characteristics:

- Is easily understood by all stakeholders
- Is briefly stated, yet clear and comprehensive in meaning
- Is challenging, yet attainable
- Is lofty, yet tangible
- Is capable of stirring excitement for all stakeholders
- Is capable of creating unity of purpose among all stakeholders
- Is not concerned with numbers
- Sets the tone for employees

From these characteristics it can be seen that crafting a worthwhile vision statement is a challenging undertaking. What follows are three vision statements—two for service providers and one for a manufacturer.

- The Institute for Corporate Competitiveness will be recognized by its customers as the provider of choice for organizational development products that are the best in the world.
- Business Express Airlines will be recognized by customers as the premiere air carrier in the United States for business travelers.
- Pendleton Manufacturing Company will be the leading producer in the United States of fireproof storage cabinets.

These vision statements illustrate the practical application of the criteria set forth earlier. Are these statements easily understood? Yes. Any stakeholder could read the vision statements and understand the dreams of the organizations they represent. Are they briefly stated, yet clear and comprehensive in meaning? Yes. Each of the statements consists of one sentence, but the sentence in each case clearly and comprehensively conveys the intended message. Are these vision statements challenging, yet attainable? Yes. Each vision presents its respective organization with the challenge of being the best in a clearly defined market and a clearly defined geographic area. Being the best in the United States or in the world is a difficult challenge in any field, but it is an attainable challenge. It can be done. Are these visions lofty, yet tangible? Yes. Trying to be the best is a lofty challenge, but, still, it is achievable and therefore tangible.

Pick a field, and some organization is going to be the best in that field. It could be this organization. Are these visions capable of stirring excitement among stakeholders? Yes. Trying to be the best in any endeavor is an exciting undertaking, the kind people want to be a part of.

Are these visions capable of creating unity of purpose? Yes. All three give stakeholders a common rallying cry. This is the type of thing that happens when a sports team sets its sights on the championship. The players, coaches, fans, and management all rally around the vision, pulling together as one in an attempt to achieve it. Do these statements concern themselves with numbers? No. Numbers are left for later in the strategic planning process. Do these visions set the tone for employees? Yes. Clearly, the organizations in question are going somewhere, and employees are expected to do their part to ensure that the organizations get there expeditiously.

DEVELOPING THE MISSION

We have just seen that the vision statement describes what an organization would like to be. It's a dream, but it's not pie in the sky. The vision represents a dream that can come true. The mission takes the next step and describes *who* the organization is, *what* it does, and *where* it is going. Figure 2–2 contains the mission statements for the three organizations introduced in the previous section.

Assess these mission statements using the three Ws—*who, what,* and *where*—as the criteria. In the first example, the ICC describes *who* it is as follows: "a business development company dedicated to helping organizations continually improve their ability to compete in the global marketplace." This description of who the ICC is also describes who its customers are. Regardless of whether both who's can be explained in one sentence, both should be explained in the mission. *What* ICC does is described as follows: "provides high-quality, competitiveness-enhancing products and services." From this statement an outsider with no knowledge of ICC could determine what the company does. *Where* ICC is going is described as reaching "an ever increasing number of organizations in the United States." Clearly, ICC wants to grow as much as possible within the geographic boundaries of the United States.

In the second example, BEA describes *who* it is as a "domestic air carrier dedicated to providing business travelers with air transportation." This simple statement describes both BEA and who its customers are. BEA is a domestic air carrier, and its customers

---

- The Institute for Corporate Competitiveness (ICC) is a business-development company dedicated to helping organizations continually improve their ability to compete in the global marketplace. To this end, ICC provides high-quality, competitiveness-enhancing products and services to an ever increasing number of organizations in the United States.
- Business Express Airlines (BEA) is a domestic air carrier dedicated to providing business travelers with air transportation that exceeds their expectations in terms of cost, convenience, service, and dependability. To this end, BEA provides air carrier service to and from a steadily increasing number of major hub airports in the United States.
- Pendleton Manufacturing Company is a hazardous-materials storage company dedicated to making your work environment safe and healthy. To this end, Pendleton produces high-quality fireproof cabinets for safely storing toxic substances and hazardous materials for an ever-broadening market in the United States.

Figure 2–2
Sample Mission Statements

are business travelers. *What* BEA does is described as "providing business travelers with air transportation that exceeds their expectations." *Where* BEA is going can be seen in the following portion of the mission statement: "BEA provides air carrier service to and from a steadily increasing number of major hub airports." Like ICC, BEA wants to grow continually in the United States.

In the third example, Pendleton Manufacturing describes *who* it is as a "hazardous materials storage equipment company dedicated to making your work environment safe and healthy." From this statement one can easily discern who Pendleton Manufacturing and its customers are. Any company that either produces hazardous waste or uses toxic materials is a potential customer. *What* Pendleton Manufacturing does is described as follows: "produces high-quality fireproof cabinets for safely storing toxic substances and hazardous wastes." *Where* Pendleton Manufacturing is going can be seen in that part of the final sentence of the mission statement that says it wants to serve "an ever broadening market in the United States."

All three of the companies in these examples want to grow continually, but only in domestic markets. No interest is expressed in international markets. This is a major strategic decision that will determine the types of actions taken to achieve their respective missions.

In developing the mission statement for any organization, one should apply the following rules of thumb:

- Describe the *who, what,* and *where* of the organization, making sure the *who* component describes the organization and its customers.
- Be brief, but comprehensive. Typically one paragraph should be sufficient to describe an organization's mission.
- Choose wording that is simple, easy to understand, and descriptive.
- Avoid *how* statements. How the mission will be accomplished is described in the "Strategies" section of the strategic plan.

DEVELOPING THE GUIDING PRINCIPLES

An organization's guiding principles establish the framework within which it will pursue its mission. Each guiding principle encompasses an important organizational value. Together, all of the guiding principles represent the organization's value system—the foundation of its corporate culture.

Freedom through control might be one such guiding principle. It is one of the corner-stones of total quality. It is a concept that applies at all levels from line employees through executive managers. It means that once parameters have been established for a given job, level, or work unit, all employees to which the parameters apply are free to operate innovatively within them. In fact, they are encouraged to be innovative and creative within established parameters. This means that as long as they observe applicable controls, employees are free to apply their knowledge, experience, and judgment in finding ways to do the job better. Once a way is established that is better than the existing way, that new procedure should become the standard throughout the organization.

An organization's guiding principles establish the parameters within which it is free to pursue its mission. These principles might be written as follows:

- XYZ Company will uphold the highest ethical standards in all of its operations.
- At XYZ Company, customer satisfaction is the highest priority.
- XYZ Company will make every effort to deliver the highest-quality products and services in the business.
- At XYZ Company, all stakeholders (customers, suppliers, and employees) will be treated as partners.

- At XYZ Company, employee input will be actively sought, carefully considered, and strategically used.
- At XYZ Company, continued improvement of products, processes, and people will be the norm.
- XYZ Company will provide employees with a safe and healthy work environment that is conducive to consistent peak performance.
- XYZ Company will be a good corporate neighbor in all communities where its facilities are located.
- XYZ Company will take all appropriate steps to protect the environment.

From this list of guiding principles, the corporate values of XYZ Company can be discerned. This company places a high priority on ethics, customer satisfaction, quality, stakeholder partnerships, employee input, continual improvement, a safe and healthy work environment, consistent peak performance, corporate citizenship, and environmental protection.

With these values clearly stated as the organization's guiding principles, employees know the parameters within which they must operate. When ethical dilemmas arise, as they inevitably will in business, employees know they are expected to do the right thing. If safety or health hazards are identified in the workplace, eliminating them will be a top priority. If employees spend their own time participating in community activities, they know it will reflect positively in their performance appraisal, because XYZ Company values corporate citizenship.

Developing guiding principles is the responsibility of an organization's executive management team. However, the recommended approach in a total quality organization is for executive managers to solicit input from all levels before finalizing the guiding principles.

DEVELOPING BROAD STRATEGIC OBJECTIVES

Broad strategic objectives translate an organization's mission into measurable terms. They represent actual targets the organization aims at and will expend energy and resources trying to achieve. Broad objectives are more specific than the mission, but they are still broad. They still fall into the realm of *what* rather than *how*. The *how* aspects of strategic planning come in the next step: developing specific tactics, projects, and activities for accomplishing broad objectives. Well-written broad organizational objectives have the following characteristics:

- Are stated broadly enough that they don't have to be continually rewritten
- Are stated specifically enough that they are measurable, but not in terms of numbers
- Are each focused on a single issue or desired outcome
- Are tied directly into the organization's mission
- Are all in accordance with the organization's guiding principles
- Clearly show what the organization wants to accomplish

In addition to having these characteristics, broad objectives apply to the overall organization, not to individual departments within the organization. In developing its broad objectives, an organization should begin with its vision and mission. A point to keep in mind is that broad strategic objectives should be written in such a way that their accomplishment will give the organization a sustainable competitive advantage in the marketplace. What follows is an organizational vision presented earlier as an example and its corresponding broad objectives:

The Institute for Corporate Competitiveness will be recognized by its customers as the provider of choice for organizational development products that are the best in the world.

The broad objectives that translate this vision into measurable terms are as follows:

1. To produce organizational development products of world-class quality that are improved continually
2. To provide organizational development services of world-class quality that are improved continually
3. To establish and maintain a world-class workforce at all levels of the organization
4. To continually increase the organization's market share for its existing products/services
5. To continually introduce new products/services to meet emerging needs in the organizational development market

Five Steps for Writing Broad Strategic Objectives

In actually writing broad objectives for an organization, the following five steps should be observed:

1. *Assemble input.*   Circulate the mission widely throughout the organization, and ask for input concerning objectives. Ask all stakeholders to answer the following question: "What do we have to accomplish as an organization in order to fulfill our mission?" Assemble all input received, summarize it, and prepare it for further review.
2. *Find the optimum input.*   Analyze the assembled input, at the same time judging how well individuals' suggestions support the organization's vision and mission. Discard those suggestions that are too narrow or that do not support the vision and mission.
3. *Resolve differences.*   Proposed objectives that remain on the list after step 2 should be discussed in greater depth in this step. Allow time for participants to resolve their differences concerning the objectives.
4. *Select the final objectives.*   Once participants have resolved their differences concerning the proposed objectives, the list is finalized. In this stage the objectives are rewritten and edited to ensure that they meet the criteria listed earlier.
5. *Publicize the objectives.*   All stakeholders need to know what the organization's objectives are. Employees, managers, suppliers, and even customers have a role to play in accomplishing the organization's objectives. These stakeholders cannot play their respective roles unless they know what the objectives are. Publicizing the organization's objectives can be done in a variety of ways. Variety and repetition are important when trying to communicate. Wall posters, wallet-sized cards, newsletters, personal letters, company-wide and departmental meetings, videotaped presentations, and annual reports can all be used to publish and communicate the organization's objectives. A rule of thumb to follow is *the more different communication vehicles used, the better.* It's also a good idea to publish the objectives along with the vision, mission, and guiding principles.

Cautions Concerning Broad Strategic Objectives

Before actually developing broad objectives for an organization, it is a good idea to become familiar with several applicable cautions. These cautions are as follows:

■ Restrict the number of objectives to just a few—from five to eight. This is a rule of thumb, not an absolute. However, if an organization needs more than eight objectives, it may be getting too specific.

- Keep the language simple so that the objectives are easily understood by all employees at all levels of the organization.
- Tie all objectives not just to the mission but also to the vision. All resources and efforts directed toward achieving the broad objectives should support the mission and the vision.
- Make sure objectives do not limit or restrict performance. This is best accomplished by avoiding numerical targets when writing them.
- Remember that achieving objectives is a means to an end, not an end in itself (the vision is the end).
- Do not use broad objectives in the employee appraisal process. The only aspect of the overall strategic plan that might be used in the employee appraisal process is the specific-tactics component. This is because only the specific tactics in the strategic plan are assigned to specific teams or individuals and given specific time frames within which they should be completed. Broad objectives, on the other hand, are everyone's responsibility.
- Relate broad objectives to all employees. This means there should be objectives covering the entire organization. Employees should be able to see that their work supports one or more of the broad objectives.
- Make broad objectives challenging, but not impossible. Good objectives will challenge an organization without being unrealistic.

DEVELOPING SPECIFIC TACTICS FOR ACCOMPLISHING THE BROAD STRATEGIC OBJECTIVES

Specific tactics are well-defined, finite projects and activities undertaken for the purpose of achieving a specific desired outcome. They are undertaken for the purpose of accomplishing an organization's broad strategic objectives. Tactics have the following characteristics:

- Are specific in nature
- Are measurable
- Can be quantifiable
- Can be accomplished within a specified time frame
- Can be assigned to a specific individual or group
- Are tied directly to a broad objective

Drafting the Individual Tactics

In drafting tactics, an organization should begin with its broad strategic objectives. Each objective will have at least one, but typically three or four, tactics accompanying it. Figure 2–3 is a tool that can be used for drafting tactics. Notice that it contains the broad strategic objective to which the tactics relate. This is not necessarily a complete list of tactics for this objective; these are just examples of tactics that might be developed in support of the objectives. The nature of tactics is such that they are accomplished and then replaced by new tactics relating to the relevant objective.

Evaluate the five tactics in Figure 2–3 by applying the criteria set forth earlier. Are these tactics specific in nature? Yes. They are finite and limited in scope. Are the tactics measurable? Yes. In each case, the organization can easily determine whether each activity was completed within the specified time frame. Are the tactics quantifiable? Only tangentially, in that one can determine whether *all* employees received the desired service within the specified time frame. It is not necessary for all tactics to be quantifiable. This is an optional criteria. Can the tactics be accomplished within a specified

---

Broad Strategic Objective
To establish and maintain a world-class workforce at all levels of the organization.

Tactic	Responsible Indiv./Unit	Timeframe/Deadline
1. Arrange TQ training for all executive managers.	CEO	Completed by January 15
2. Arrange teamwork training for all executive managers.	CEO	Completed by January 30
3. Give all employees training in the use of problem-solving/quality tools.	Department Managers	January 15–February 20
4. Give all employees training in continual improvement methods.	Department Managers	March 15–May 15
5. Establish a company-supported off-duty education program for all employees.	Human Resources Department	In place by February 28

Figure 2–3
Tactics Development Form

time frame? Yes. A time frame is specified for the completion of each tactic. Can the tactics be assigned to a specific individual or group? Yes. In each case, a responsible party is named by position. Do the tactics tie directly to a specific broad objective? Yes. The related objective is shown on the form in the example. All of the tactics meet the applicable criteria.

EXECUTING THE STRATEGIC PLAN

The old saying "The best-laid plans of mice and men often go astray" is, unfortunately, all too true. Many organizations devote time, energy, and money to developing comprehensive, thorough, detailed plans, only to see them come apart at the seams shortly after they begin to be executed. Execution is a critical component of strategic management, but for some reason it rarely receives the attention it deserves. To put less energy and thought into execution than into planning is a major error because even the best strategic plan won't help an organization if it is poorly executed.

Picture the following scenario. A family plans a two-week vacation to a national park. The parents envision a relaxing, fun-filled two weeks of camping, hiking, swimming, and biking. Their mission (that of the parents) is to get away together and share some quality time as a family. Certain guiding principles concerning behavior, spending, and work sharing are established by the parents. They also establish some broad objectives concerning the various activities the family wants to pursue. Wisely, the parents involve the children in this step. Also working with the children, parents draft strategies for accomplishing their objectives.

The family had an excellent plan for an enjoyable vacation, but as soon as the plan went into execution, the family began to have problems:

- Disagreements among the children concerning destinations and activities
- Setup problems when the children did not know how to perform some of their assigned duties (e.g., setting up the tent at the campsite, building a proper campfire, monitoring daily gas mileage while the parents drove)

■ Attitude problems concerning various aspects of the trip, including distance to cover daily, how many rest stops to make, where to eat while driving to the eventual destination

As a result of these problems, the planned vacation of shared family fun and relaxation turned into an emotionally draining two weeks of stress, anger, and frustration. The family's problems were the results of faulty execution of the vacation plan. The parents in this example failed to apply the following steps, all of which are critical to successful execution:

■ *Communicate.* Make sure all stakeholders understand the plan and where they fit into it.

■ *Build capabilities.* Make sure all stakeholders have the skills needed to carry out their assignments and responsibilities in the plan.

■ *Establish strategy-supportive stimuli.* People in the workplace respond to stimuli. When trying to execute a plan, it is important to ensure that strategy-supportive stimuli are in place. Typically, in a work setting the most effective stimuli are reward and recognition incentives. It is not uncommon to find that an organization's strategic plan expects people to move in one direction, while its incentives encourage them to move in another.

■ *Eliminate administrative barriers.* Every organization establishes administrative procedures for accomplishing its day-to-day work. If executing a new strategy changes the intended direction of the organization, administrative procedures may need to be changed correspondingly. A mistake commonly made by organizations attempting to execute a plan is leaving outdated administrative procedures in place. Administrative procedures put in place when the organization was moving in one direction can become inhibitors when the organization decides to move in another direction.

■ *Identify advocates and resisters.* In any organization there will be advocates and resisters when it comes to executing the strategic plan. This is natural and should be expected. As the plan unfolds, if it is successful, resisters will become advocates. If it fails, advocates will try to distance themselves from the plan, and resisters will say, "I told you so." This is just human nature and should be expected. Consequently, it is important to give the plan the best possible chance of succeeding. One way to do this is to assign all initial activities to advocates. Giving initial assignments to resisters when executing a new strategic plan is likely to ensure failure of the plan. Eventually all employees must play a role in executing the strategic plan, but in the critical early stages, stick with advocates and avoid resisters.

■ *Exercise strategic leadership.* It is important that managers at all levels set a positive example by (a) showing that they believe in the strategic plan, (b) ensuring that all decisions are based on that action that best supports the strategic plan and (c) allocating resources based on priorities established in the strategic plan.

■ *Be flexible and improvise.* According to General Colin Powell, "No battle plan survives contact with the enemy."[1] General Powell's statement could easily be misconstrued. After all, if he is right, why should an organization go to the trouble of planning in the first place? Powell's statement does not advocate against planning. Rather, it advocates flexibility. Plans, when they are developed, are based on assumptions that may not be accurate. They are also based on a presumed set of circumstances, circumstances that even if they were accurate when the plan was developed might change before the plan is executed. This is why strategic plans must be viewed as a set of flexible guidelines rather than a hard and fast road map from which it is impossible to deviate. Every traveler knows there will be detours on even the best planned trip. Managers who want to make their organizations competitive must be willing to improvise when necessary.

During the 1970s, teachers of grades K–12 had to learn how to develop lesson plans. They were taught by seminar leaders and college professors to plan their daily lessons right down to the minute, with each activity assigned a specific amount of time. Of course, the wiser, more experienced teachers were able to predict the outcome of this approach, but sometimes it is best to let reality make your point for you. As teachers attempted to hold rigidly to schedules built into their plans, they found themselves falling farther and farther behind as problems cropped up they had not foreseen when developing their lesson plans.

Before long, lesson plans were scrapped and replaced by planned lessons. The difference between a lesson plan and a planned lesson can be summed up in one word: *flexibility*. Planned lessons give teachers a general direction, some expected outcomes, and loose guidelines on how to get where they are going. They contain no rigid time constraints but, more important, give teachers the flexibility to change directions and pursue a whole new set of activities if an opportunity for learning presents itself.

This example holds a lesson for managers who develop and execute strategic plans. Be flexible. The assumptions on which the plan was built might not be accurate. The circumstances in which the plan was supposed to be implemented might change. Planning is necessary so that resources can be properly allocated and so that employees on whom the organization depends for progress can get a better picture of where the organization is trying to go. But if somebody moves the target, don't continue to shoot at the spot where it used to be.

This gets back to the point made at the beginning of this section by General Powell's quote. Organizations should plan thoroughly and carefully, based on the most accurate assumptions possible at a given point in time. However, if upon implementing the plan, it becomes obvious that the assumptions are not valid or that circumstances have changed, organizations should not rigidly adhere to steps that no longer make sense. If the plan, or any part of it, is no longer valid, improvise and move on.

■ *Monitor and adjust as needed.*   Developing and executing a strategic plan is an example of the *plan-do-check-adjust* cycle in action. The *plan* component of the cycle involves developing the strategic plan. The *do* component is the execution phase. The *check-adjust* component involves monitoring progress toward completion of specific strategies and making the necessary adjustments when roadblocks are encountered. Will it take longer than you thought to complete a project or activity? Adjust the time frame. Have unexpected barriers been encountered? Decide what needs to be done to overcome the barriers, and do it. Did you complete the project only to find it didn't yield the expected results? Develop a substitute tactic and try again.

===== ENDNOTES =====

1. Colin Powell, as quoted in "Good/Bad News about Strategy" by Oren Harari, *Management Review* 84, no. 7 (July 1995): 29.

CHAPTER THREE

Quality Culture

- Understanding What a Quality Culture Is
- Activating Cultural Change
- Laying the Groundwork for a Quality Culture
- Learning What a Quality Culture Looks Like
- Countering Resistance to Cultural Change
- Establishing a Quality Culture

One of the greatest obstacles faced by organizations attempting to implement total quality is the cultural barrier. Many organizations do an excellent job of committing to total quality, involving employees in all aspects of planning and implementation, and providing the training needed to ensure that employees have the necessary skills, only to have their efforts fall flat. The culprit in many of these cases is organizational inertia. No effort has been made to overcome the comfort employees at all levels feel in doing things the way they have always been done. In other words, no effort has been made to change the organization's culture. This chapter explains the concept of organizational culture and how to go about changing it.

UNDERSTANDING WHAT A QUALITY CULTURE IS

To understand what a quality culture is one must first understand the concept of *organizational culture*. Every organization has one. An organization's culture is the everyday manifestation of its underlying values and traditions. It shows up in how employees behave on the job, their expectations of the organization and each other, and what is considered normal in terms of how employees approach their jobs. Have you ever shopped at a store or eaten in a restaurant in which the service was poor and the employees surly or disinterested? Such organizations have a cultural problem. Valuing the customer is not part of their culture. No matter what slogans or what advertising gimmicks they use, the behavior of their employees clearly says, "We don't care about customers."

An organization's culture has the following elements:

- Business environment
- Organizational values
- Cultural role models
- Organizational rites, rituals, and customs
- Cultural transmitters

The business environment in which an organization must operate is a critical determinant of its culture. Organizations that operate in a highly competitive business environment that changes rapidly and continually are likely to develop a *change-oriented* culture. Organizations that operate in a stable market in which competition is limited may develop a *don't-rock-the-boat* culture.

Organizational values describe what the organization thinks is important. Adherence to these values is synonymous with success. Consequently, an organization's values are the heart and soul of its culture.

Cultural role models are employees at any level who personify the organization's values. When cultural role models retire or die, they typically become legends in their organizations. While still active, they serve as living examples of what the organization wants its employees to be.

Organizational rites, rituals, and customs express the organization's unwritten rules about how things are done. How employees dress, interact with each other, and approach their work are all part of this element of an organization's culture. Rites, rituals, and customs are enforced most effectively by peer pressure.

Cultural transmitters are the vehicles by which an organization's culture is passed down through successive generations of employees. The grapevine in any organization is a cultural transmitter, as are an organization's symbols, slogans, and recognition ceremonies.

What an organization truly values will show up in the behavior of its employees, and no amount of lip service or advertising to the contrary will change this. If an organization's culture is its value system as manifested in organizational behavior, what is a quality culture?

> A quality culture is an organizational value system that results in an environment that is conducive to the establishment and continual improvement of quality. It consists of values, traditions, procedures, and expectations that promote quality.

How do you recognize an organization with a quality culture? It is actually easier to recognize a quality culture than to define one. Organizations with a quality culture, regardless of the products or services they provide, share a number of common characteristics, presented in Figure 3–1.

✓ Behavior matches slogans.

✓ Customer input is actively-sought and used to continually improve quality.

✓ Employees are both involved and empowered.

✓ Work is done in teams.

✓ Executive-level managers are both committed and involved; responsibility for quality is *NOT* delegated.

✓ Sufficient resources are made available where and when they are needed to ensure the continuous improvement of quality.

✓ Education and training are provided to ensure that employees at all levels have the knowledge and skills needed to continuously improve quality.

✓ Reward and promotion systems are based on contributions to the continual improvement of quality.

✓ Fellow employees are viewed as internal customers.

✓ Suppliers are treated as partners.

Figure 3–1

Characteristics Shared by Organizations with a Quality Culture

How Are Organizational Cultures Created?

Many factors contribute to the creation of an organization's culture. The value systems of executive-level decision makers are often reflected in their organization's culture. How managers treat employees and how employees at all levels interact on a personal basis also contribute to the organizational culture. Expectations are important determinants of organizational culture. What management expects of employees and what employees, in turn, expect of management both contribute to an organization's culture. The stories passed along from employee to employee typically play a major role in the establishment and perpetuation of an organization's culture. All of these factors can either help or hurt an organization.

If managers treat employees with trust, dignity, and respect, employees will be more likely to treat each other in this way, and trust, dignity, and respect in everyday interaction will become part of the organization's culture. On the other hand, if management treats employees poorly, employees are likely to follow suit. Both situations, if not changed, will become ingrained as traditions. These traditions will be perpetuated both by the behavior of employees and by the stories they pass along to one another. This is why it is so important to establish a quality culture. If mistrust is part of the organizational culture, it will be difficult to build partnerships between internal and external customers. It will also be difficult to establish an environment of mutually supportive teamwork.

Commitment to quality cannot be faked. Employees know when management is just going through the motions. Changing an organization's culture requires a total commitment and a sustained effort at all levels of the organization.

ACTIVATING CULTURAL CHANGE

To attempt the implementation of total quality without creating a quality culture is to invite failure. Organizations in which the prevailing culture is based on traditional management practices are not likely to succeed in the implementation of total quality. Successful total quality requires cultural change. Several primary reasons cultural change must either precede or at least parallel the implementation of total quality are described here.

Change Cannot Occur in a Hostile Environment The total quality approach to doing business may be radically different from what management and employees are used to. Managers who are used to sitting in their lonely towers at the top of the pecking order and issuing edicts from on high are likely to reject the concept of employee involvement and empowerment.

Employees who are used to competing against their fellow employees for promotions and wage increases may not be open to mutually supportive internal partnerships and teamwork. Situations such as these can create an environment that is hostile toward change, no matter how desirable that change is. Change can be difficult, even when people want to change. It can be impossible in a hostile environment.

Moving to Total Quality Takes Time The nature of total quality is such that the organization may have to go down somewhat before it can turn things around and start to come up. In a conversion to total quality, positive results are rarely achieved in the short run. This characteristic gives nonbelievers and people who just don't want to change (and such people are often in the majority at first) the opportunity to promote the "I told you it wouldn't work syndrome."

It Can Be Difficult to Overcome the Past Employees who have worked in an organization for any period of time have probably seen a variety of management fads come and go. Promoting the latest management gimmick and then letting it die for lack of interest may be part of the existing organizational culture. If this is the case, it will be difficult to overcome the past. Employees will remember earlier fads and gimmicks and characterize total quality as being just the latest of these and take a "This too shall pass" attitude toward it. The past is not just an important part of an organization's culture; it can be the most difficult part to overcome.

LAYING THE GROUNDWORK FOR A QUALITY CULTURE

Establishing a quality culture is a lot like constructing a building. First, you must lay the groundwork, or foundation. According to Peter Scholtes, this should begin by developing an understanding of what he calls the "laws" of organizational change.[1] These laws are explained in the following paragraphs.[2]

Understand the History Behind the Current Culture   Organizational cultures don't just happen. Somebody wrote the policy that now inhibits competitiveness. Somebody started the tradition that is now such a barrier. Times and circumstances change. Don't be too quick to criticize. Policies, traditions, and other aspects of the existing culture that now seem questionable may have been put in place for good reason in another time and under different circumstances. Learn the history behind the existing culture before trying to change it.

Don't Tamper with Systems—Improve Them   Tampering with existing systems is not the same as improving them. Tampering occurs when changes are made without understanding why a given system works the way it does and without fully understanding what needs to be changed and why. To improve something, you must first understand what is wrong with it, why, and how to go about changing it for the better.

Be Prepared to Listen and Observe   People are the primary inhibitors of change in any organization. Consequently, it is easy to become frustrated and adopt an attitude of "We could get a lot done if it weren't for the people in this organization." The problem with such an attitude is that people *are* the organization. For this reason, it is important to pay attention to both people and systems. Be prepared to listen and observe. Try to hear what is being said and observe what is not being said. Employees who are listened to are more likely to participate in changes than those who are not.

Involve Everyone Affected by Change in Making It   People will resist change. To do so is normal human behavior. What people really don't like is being changed. It can be difficult to effect change even when people want to change. It can be impossible when people feel that changes are being imposed on them. The most effective way to ensure that employees will go along with changes is to involve them in planning and implementing the changes. Give them opportunities to express their concerns and fears. Getting problems into the open from the outset will allow them to be dealt with forthrightly and overcome. Shoving them aside or ignoring them will guarantee that even little problems become big ones.

LEARNING WHAT A QUALITY CULTURE LOOKS LIKE

Part of laying the groundwork for a quality culture is understanding what one looks like. This is a lot like a person who wants to lose weight taping a picture to the mirror of someone whom that person wants to look like. The picture serves not only as a constant reminder of the destination but also as a measurement device that indicates when a goal has been met. If a picture of a company with a strong quality culture could be taped to an organization's wall for all employees too see, it would have the following characteristics:[3]

- Widely shared philosophy of management
- Emphasis on the importance of human resources to the organization
- Ceremonies to celebrate organizational events
- Recognition and rewards for successful employees
- Effective internal network for communicating the culture
- Informal rules of behavior
- Strong value system
- High standards for performance
- Definite organizational character

Knowing the laws of organizational change and understanding the characteristics of organizations that have strong quality cultures is important to any executive team that hopes to change the culture of its organization. Before implementing any of the specific strategies for establishing a quality culture that are explained later in this chapter, every person who will be involved in the change or affected by it should be familiar with these laws and characteristics.

COUNTERING RESISTANCE TO CULTURAL CHANGE

Change is resisted in any organization. Resistance to change is normal organizational behavior. In this regard, an organization is similar to a biological organism. From the perspective of organizational culture, the alien is change, and the organism is the organization to be changed. Continuous improvement means continuous change. To ensure continuous improvement one must be able to facilitate continuous change.

Why Change Is Difficult

Most people understand and accept that organizational change will be resisted. However, to be an effective agent of change, one must understand why it is resisted. Joseph Juran describes organizational change as a "clash between cultures."[4]

Advocates focus on the anticipated benefits of the change. Resisters, on the other hand, focus on perceived threats to their status, beliefs, habits, and security. Often, both advocates and resisters are wrong in how they initially approach change. Advocates are often guilty of focusing so intently on benefits that they fail to take into account the perceptions of employees who may feel threatened by the change. Resisters are often guilty of focusing so intently on threats to the status quo that they refuse to acknowledge the benefits. These approaches typically divide an organization into warring camps that waste energy and time instead of focusing resources on the facilitation of change.

How to Facilitate Change

The responsibility for facilitating change necessarily falls to its advocates. The broad steps in facilitating change are as follows:

1. Begin with a new advocacy paradigm.
2. Understand concerns of potential resisters.
3. Implement change-promoting strategies.

Begin with a New Advocacy Paradigm

The first step in facilitating change is to adopt a facilitating paradigm. Juran summarizes the traditional paradigm of change advocates as follows:[5]

- Advocates of change tend to focus solely on expected results and benefits.
- Advocates are often unaware of how a proposed change will be perceived by potential resisters.
- Advocates are often impatient with the concerns of resisters.

If change is to happen, advocates must begin with a different paradigm. When a change is advocated, ask such questions as the following:

- Who will be affected by this change and how?
- How will the change be perceived by those it affects?
- How can the concerns of those affected be alleviated?

Understand Concerns of Potential Resisters

The second step in facilitating change is to understand the concerns of potential resisters—to put yourself in their place. Philip E. Atkinson suggests that people resist change for the following reasons:[6]

- *Fear.* Change brings with it the unwanted specter of the unknown, and people fear the unknown. Worst-case scenarios are assumed and compounded by rumors. In this way, fear tends to feed on itself, growing with time.
- *Loss of control.* People value having a sense of control over their lives. There is security in control. Change can threaten this sense of security and cause people to feel as if they are losing control of their lives, jobs, areas of responsibility, and so on.
- *Uncertainty.* Uncertainty is difficult to deal with. For better or worse, people like to know where they stand. Will I be able to handle this? What will happen to me if I can't? These are the types of questions people have when confronted with change.
- *More work.* Change sometimes means more work, at least at first. This concern includes work in the form of learning. To make the change, people may have to learn more information or develop new skills. For an undefined period, they may have to work longer hours.

Implement Change-Promoting Strategies

The third step in facilitating change is implementing change-promoting strategies. These are strategies that require an advocacy paradigm and take into account the concerns people typically have when confronted with change. Juran recommends the following strategies for handling and overcoming resistance to change.[7]

Involve Potential Resisters

At some point in the process, those affected by change (potential resisters) will have to take ownership of the change or it will fail. By involving them from the outset in planning for the change, organizations can ensure that potential resisters understand it and have adequate opportunities to express their views and concerns about it. This type of involvement will help potential resisters develop a sense of ownership in the change that can, in turn, convert them to advocates.

Avoid Surprises

Predictability is important to people. This is one of the reasons they resist change. Change is unpredictable. It brings with it the specter of the unknown. For this reason, it is better to bring potential resisters into the process from the outset. Surprising potential resisters will turn them into committed resisters.

Move Slowly at First

To gain the support of potential resisters, it is necessary to let them evaluate the proposed change, express their concerns, weigh the expected benefits, and find ways to alleviate problems. This can take time. However, if advocates are perceived as rushing the change through, potential resisters will become distrustful and dig in their heels.

Start Small and Be Flexible

Change will be more readily accepted if advocates start small and are flexible enough to revise strategies that are not working as planned. This approach offers several benefits, including the following:

1. Starting with a small pilot test or experiment is less threatening than a broad-based, all-encompassing implementation.
2. Conducting a small pilot test can help identify unanticipated problems with the change.
3. The results of a pilot test can be used to revise the plans for change so that valuable resources are not wasted moving in the wrong direction.

Create a Positive Environment

The environment in which change takes place is determined by reward and recognition systems and examples set by managers. A reward and recognition system that does not reward risk taking or that punishes employees for ideas that don't work will undermine change. Managers that take "Do as I say, not as I do" attitudes will also undermine change. Well-thought-out, sincere attempts to make improvements should be recognized and rewarded even when they fail. Managers should roll up their sleeves and do their share of the work associated with change. This approach will create a positive environment that is conducive to change.

Incorporate the Change

Change will be more readily accepted if it can be incorporated into the existing organizational culture. Of course, this is not always possible. However, when it can be done, it should be done. An example might be using an established equipment maintenance schedule to make major new equipment adaptations (e.g., retrofitting manually controlled machine tools for numerical control).

Provide a Quid Pro Quo

This strategy could also be called *require something, give something*. If, for example, change will require intense extra effort on the part of selected employees for a given period of time, offer these employees some paid time off either before or immediately after the change is implemented. Using a quid pro quo can show employees that they are valued.

Respond Quickly and Positively

When potential resisters raise questions or express concerns, advocates should respond quickly and positively. Making employees wait for answers magnifies the intensity of their concerns. A quick response can often eliminate the concern before it becomes a problem, and it will show employees that their concerns are considered important. A quick response does not mean a surface-level or inaccurate response made before having all the facts. Rather, it means a response made as soon as it can be made thoroughly and accurately. It is also important to respond positively. Advocates should not be offended by or impatient with the questions of potential resisters. A negative attitude toward questions and concerns only serves to magnify them.

Work with Established Leaders

In any organization, some people are regarded as leaders. In some cases, those people are in leadership positions (supervisors, middle managers, team captains, etc.). In other cases, they are informal leaders (highly respected employees whose status is based on their experience or superior knowledge and skills). The support of such leaders is critical. Other employees will take cues from them. The best way to get their support is to involve them in planning for the change from the outset.

Treat People with Dignity and Respect

This strategy is fundamental to all aspects of total quality. It requires behavior that acknowledges the human resource as the organization's most valuable resource. Without this strategy, the others won't matter.

Be Constructive

Change is not made simply for the sake of change. It is made for the sake of continual improvement. Consequently, it should be broached constructively from the perspective of how it will bring about improvements.

ESTABLISHING A QUALITY CULTURE

Establishing a quality culture involves specific planning and activities for every business or department. This section identifies the steps involved, but first it outlines the emotional processes employees go through as the steps are being taken. Managers need to recognize and accommodate the emotional transition required not only of employees but also of themselves while the steps toward making the conversion to quality go on.

Phases of Emotional Transition

A great deal of research has been done about how people undergo transitions from one state of being to another. Most of this research has focused on the stages of transition or recovery that people go through when they confront a major unexpected and unwanted change in their lives. The types of changes that have been studied most include divorce, the death of a loved one, a life-threatening illness, and the loss of a job.

The first emotional response to any type of change is shock. A person is living from day to day, comfortable with the predictability of his or her life. Suddenly an unexpected change intrudes. A typical response to the shock it produces is denial. The change is so unwanted that the natural human response is to simply deny that it has happened. This serves to level the state of mind somewhat from the low experienced during the shock phase. The length of the denial phase differs from person to person. Regardless of its length, the denial phase is temporary.

Events force the issue, and the realization of reality begins to set in. As this happens, the person's state of mind begins to fall. Depression is common during the realization phase. People need a lot of support during this phase. When realization bottoms out, acceptance occurs. Acceptance does not mean the person agrees with what has happened. Rather, it means that he or she is ready to say, "I have this problem; now what can I do about it?"

This attitude allows the rebuilding process to begin. During this phase, people need as much support as they did during the realization phase. As the rebuilding phase is accomplished, understanding sets in. In this phase, people have come to grips with the change, and they are dealing with it successfully. This phase blends into the final phase, recovery. In this phase, people are getting on with their lives.

Managers hoping to instill a quality culture should understand this transitional process. The change from a traditional organizational culture to a quality culture can

> ✓ Identify the attitudes, behaviors, processes, and procedures that are to be changed.
> ✓ Put the planned changes in writing.
> ✓ Develop a comprehensive plan for making the changes.
> ✓ Make sure all change advocates are familiar with the emotional transition people go through when confronted with change.
> ✓ Identify the key people in the organization who can either make the conversion work or make sure it doesn't work.
> ✓ Get the identified key people on the team (turn them into advocates).
> ✓ Take a *hearts and minds* approach when introducing the new culture.
> ✓ Apply *courtship* strategies to bring people along slowly but steadily.
> ✓ SUPPORT, SUPPORT, SUPPORT.

Figure 3–2
Quality Culture Conversion Checklist

be traumatic enough to trigger the process. Knowing this and understanding the process will help managers who are trying to instill a quality culture.

Steps in the Conversion to Quality

Figure 3–2 provides a checklist managers can use to guide their organizations through the conversion to a quality culture. The various strategies contained in the checklist are explained in the following subsections.

Identify the Changes Needed

An organization's culture dictates how people in it behave, respond to problems, and interact with each other. If the existing culture is a quality culture, it will have such characteristics as the following:

- Open, continual communication
- Mutually supportive internal partnerships
- Teamwork approach to problems and processes
- Obsession with continual improvement
- Broad-based employee involvement and empowerment
- Sincere desire for customer input and feedback

Does the organization's culture have these characteristics? The best way to answer this question is to involve the entire workforce from bottom to top in a systematic assessment that is stratified by level (i.e., executive management, middle management, first-line employee, etc.). Figure 3–3 is an example of an assessment instrument that can be used for collecting information on the perceptions of employees at all levels in an organization.

Put the Planned Changes in Writing

A comprehensive assessment of an organization's existing culture will usually identify improvements that need to be made. These improvements will require changes in the status quo. These changes should be listed without annotation or explanation. For example, if the assessment reveals that customer input is not part of the product development cycle, the change list would contain an entry such as the following: *The product development cycle should be changed so that it includes the collection and use of customer input.*

Position (Type) _____

Date _____

Instructions

    The purpose of this survey is to assess the existing culture of organization. The findings will be compared with what is known about a quality culture for the purpose of identifying the cultural changes needed in our organization to continually improve quality, productivity, and competitiveness. Respond to each of the criteria by circling the number you think best describes our organization as it is today. *Zero* (0) means that we do not meet this criterion at all. *Five* (5) means that we completely satisfy the criterion. Do not respond to items that don't apply or about which you are unsure.

1. All employees know the mission of the organization   0 1 2 3 4 5
2. All employees know their role in helping the organization accomplish its mission   0 1 2 3 4 5
3. Executive management is committed to the continual improvement of quality, productivity, and competitiveness   0 1 2 3 4 5
4. Management treats the workforce as a valuable asset   0 1 2 3 4 5
5. Open, continual communication exists at all levels of the organization   0 1 2 3 4 5
6. Mutually supportive internal partnerships exist between management and employees   0 1 2 3 4 5
7. Mutually supportive internal partnerships exist among employees   0 1 2 3 4 5
8. Quality is defined by customers, internal and external   0 1 2 3 4 5
9. Customers participate in the product development cycle   0 1 2 3 4 5
10. Employees are involved in the decision-making process   0 1 2 3 4 5
11. Employees are empowered to contribute their ideas for promoting continual improvement   0 1 2 3 4 5
12. Performance of processes is measured scientifically   0 1 2 3 4 5
13. Scientific data are used in the decision-making process   0 1 2 3 4 5
14. Employees receive the education and training they need to continually improve their performance   0 1 2 3 4 5
15. All employees at all levels are expected to maintain high ethical standards   0 1 2 3 4 5

Figure 3–3
Organizational Culture Employee Assessment Worksheet

Develop a Plan for Making the Changes

The plan for effecting change is developed according to the Who-What-When-Where-How model. Each of these elements represents a major section of the plan, as follows:

- Who will be affected by the change? Who will have to be involved in order for the change to succeed? Who is likely to challenge the change?

- What tasks must be accomplished? What are the most likely barriers? What are the related processes and procedures that will be affected by the change?

- When should the change be implemented? When should progress be measured? When should the various tasks associated with the change be accomplished? When should implementation be completed?

- Where will the change be implemented? Where are the people and processes that will be affected?

■ How should the change be made? How will it affect existing people and processes? How will it improve quality, productivity, and competitiveness?

The plan should contain all five elements, and each element should be comprehensively dealt with. However, the plan should be brief. Be comprehensive and thorough, but keep it as brief as possible.

Understand the Emotional Transition Process

Advocates of the change will play key roles in its implementation. The success of the implementation will depend to a large extent on how well advocates play their roles. It is essential that they understand the emotional transition people go through when forced to deal with change, particularly unwanted change.

The transition consists of seven steps: shock, denial, realization, acceptance, rebuilding, understanding, and recovery. People who confront a change they don't want to make may have to go through all seven steps in the transition. Advocates should understand this and proceed accordingly.

Identify Key People and Make Them Advocates

Key people are those who can facilitate and those who can inhibit implementation of the change. These people should be identified, brought together, and given the plan. Give advocates and inhibitors opportunities to state their cases. Record all concerns and deal with them. This is the step in which a quid pro quo might be used to bring inhibitors around. Executive managers must use their judgment in applying the right amount of the carrot, the stick, and peer pressure (from advocates) to turn inhibitors into advocates.

Take a Hearts and Minds Approach

Advocates should be conscious of human nature as they work to implement change. On an intellectual level, people may understand and even agree with the reasons behind a change. But understanding intellectually is rarely enough. People tend to react to change more on an emotional (hearts) level than on an intellectual (minds) level, at least initially. Therefore, it is important to take the time to deal with the inevitable emotional response that occurs in the early stages of implementation.

Frequent, open communication—preferably face-to-face—is the best strategy. Advocates should allow even the most negative opponents to voice their concerns and objections in open forums. Then these concerns should be answered in an objective, patient, nondefensive manner. When the majority of employees accept the change, critical mass will set in and peer pressure will begin to work on the side of the advocates.

Apply Courtship Strategies

Courtship is a phase in a relationship that moves slowly but deliberately toward a desired end. During the courtship, the partner hoping to move the relationship forward listens carefully to the other partner and patiently responds to any concerns expressed. This partner is on his or her best behavior. If advocates think of their relationship with potential resisters as a courtship, they will be better able to bring them along and eventually win them over.

Support, Support, Support

This final strategy is critical. It means that the material, moral, and emotional support needed by people undergoing change should be provided. Undergoing change is a lot like walking a tightrope for the first time. It will work out a lot better if you have someone to help you get started, someone waiting at the other end to encourage progress, and a safety net underneath in case you fall. Planning is important. Communication is critical. But support is essential.

========= ENDNOTES =========

1. Peter R. Scholtes, The Team Handbook (Madison, WI: Joiner Associates, 1992), 1–20.
2. Scholtes, 1-20–1-21.
3. Joseph M. Juran, *Juran on Leadership for Quality* (New York: Free Press, 1989), 184.
4. Juran, 316.
5. Juran, 317.
6. Philip E. Atkinson, *Creating Culture Change: The Key to Successful Total Quality Management* (San Diego, CA: Pfeiffer, 1990), 48–49.
7. Juran, 318–319.

Customer Satisfaction

In a total quality setting, customers define quality and employees produce it. Historically, organizations have viewed customers as people who buy and use their products. These are external customers. There are also internal customers within any organization—the staff. With this background, an accurate recasting of the first sentence is as follows: In a total quality setting, external customers define quality and internal customers produce it. This chapter provides the information modern managers need to establish in their organizations a customer focus that encompasses both internal and external customers.

UNDERSTANDING CUSTOMER-DEFINED QUALITY

In a total quality setting, quality is defined by the customer. According to Peter R. Scholtes:

> Only once you understand what processes and customers are will you be able to appreciate what quality means in the new business world. If customers are people who receive your work, only they can determine what quality is, only they can tell you what they want and how they want it. That's why a popular slogan of the quality movement is "quality begins with the customer." You must work with internal and external customers to determine their needs, and collaborate with internal and external suppliers.[1]

In his book *Total Manufacturing Management*, Giorgio Merli makes the following points about customer-defined quality:[2]

- The customer must be the organization's top priority. The organization's survival depends on the customer.
- Reliable customers are the most important customers. A reliable customer is one who buys repeatedly from the same organization. Customers who are satisfied with the quality of their purchases from an organization become reliable customers. Therefore, customer satisfaction is essential.

■ Customer satisfaction is ensured by producing high-quality products. It must be renewed with every new purchase. This cannot be accomplished if quality, even though it is high, is static. Satisfaction implies continual improvement. Continual improvement is the only way to keep customers satisfied and loyal.

If customer satisfaction is the highest priority of a total quality organization, then it follows that such an organization must have a customer focus. Traditional management practices that take the management by results approach are inward looking. An organization with a customer focus is outward looking. Scholtes describes this concept as follows:

> Whereas Management by Results begins with profit and loss and return on investment, Quality Leadership begins with the customer. Under Quality Leadership, an organization's goal is to meet and exceed customer needs, to give lasting value to the customer.[3]

The key to establishing a customer focus is putting employees in touch with customers and empowering those employees to act as necessary to satisfy the customers. There are a number of ways to put employees in touch with customers. Actual contact may be in person, by telephone, or through reviewing customer-provided data. Identifying customer needs and communicating with customers are covered later in this chapter. At this point, it is necessary to understand only that employee–customer interaction is a critical element in establishing a customer focus.

An excellent example of how a company can establish a customer focus is Zytec Corporation of Eden Prairie, Minnesota. Zytec manufactures power supplies. Writing about Zytec's innovative customer focus, the *Total Quality Newsletter* said this:

> At the heart of its corporate commitment to customer satisfaction, for example, is this message to each of its 700 employees: To solve problems for our customers, you are empowered to spend up to $1,000 on their behalf. No questions asked. There is no paperwork or authorization needed to spend money for a customer. Employees can requisition a check or— if more immediate action is necessary—pay cash for the purchase and be reimbursed. What kind of actions have Zytec employees taken for customers? One chartered a plane to fly some missing equipment pieces from a Zytec plant in southeastern Minnesota to a customer in a central part of the state; another bought new tables for a conference room where customers frequently visited.[4]

IDENTIFYING CUSTOMER NEEDS

Historically, customers were excluded from the product development process. When this approach is used, the organization producing the product is taking a chance that it will satisfy the customer. In a competitive marketplace that is global in scope, such an approach can be disastrous. In a total-quality setting, customer needs are identified clearly as a normal part of product development. According to Scholtes:

> The goal should be to exceed customer expectations, not merely meet them. Your customers should boast about how much they benefit from what you do for them. To attain this goal, you must collect reliable information on what they need and want from your product or service. In doing so, you will find out whether your processes are on target. This strategy can be used to identify potential improvement projects or just to clarify a project's goals.[5]

Scholtes recommends the following six-step strategy for identifying customer needs:[6]

Speculate about Results

Before gathering information about customer needs, it is a good idea to spend some time speculating about what might be learned. Write down what you think customers will say, so that you can compare your expectations with what is actually said. The purpose of this step is to help representatives of the organization determine whether they are in touch with customer needs.

Develop an Information-Gathering Plan

Information gathering should be systematically undertaken and well organized. Before gathering information, develop a plan. Decide what types of information are needed and who will be asked to provide it. Whenever possible, structure the plan so that information is collected in face-to-face interviews. When personal visits are not possible, use the telephone. Written surveys sent out through the mail cannot produce a level of feedback equal to that gained from the nonverbal messages, impressions, and follow-up questions that are a part of person-to-person interviews.

Gather Information

Before implementing the entire information-gathering plan, it is a good idea to conduct a smaller pilot study involving just a few customers. This will identify problems with the information-gathering methodology that should be corrected before you proceed on a larger scale. After the methodology has been appropriately refined, gather information in a timely manner.

Analyze the Results

Results should be analyzed carefully and objectively. Do they match the speculated results from the first step? How do they agree and disagree? What problems did customers identify? What strong points? Were there trends? How many customers complained of the same problem? What changes in the product or services relating to it were suggested?

Check the Validity of Conclusions

Having drawn conclusions based on the information gathered, the next step is to check the validity of those conclusions. Customers can be a valuable source of help. Select several customers and share the conclusions with them. Do they agree with the conclusions? Also share the conclusions with other people in the organization and get their feedback. Adjust your conclusions as needed based on this external and internal feedback.

Take Action

Based on the final conclusions, what changes need to be made? Which of these changes are short-term in nature, and which are long-term? Which can be made immediately, and which will require longer? Take any corrective action that can be done immediately, and lay out a plan for completing any that is long-term in nature. Meet with customers and let them know what is going to be done and when. Make sure that changes are made, to the extent possible, in the same order of priority as that dictated by customer needs.

COMMUNICATING WITH CUSTOMERS

Continual communication with customers is essential in a competitive marketplace. Establishing effective mechanisms for facilitating communication and then making sure the mechanisms are used are critical strategies in establishing a customer focus. One of the main reasons continual communication is required is that customer needs change, and, at times, they can change rapidly. Quality pioneer Joseph M. Juran explains this concept as follows:

> Customers' needs do not remain static. There is no such thing as a permanent list of customers' needs. We are beset by powerful forces that keep coming over the horizon and are

ever changing directions: new technology, market competition, social upheavals, international conflicts. These changing forces create new customer needs or change the priority given to existing ones.[7]

Communication with customers must extend to both external and internal customers. What applies on the outside also applies within the organization. According to Juran:

> Most companies do a good job of communicating essential quality information to workers on matters such as specifications, standards, procedures, and methods. This information, although necessary, is not sufficient with respect to willful errors. We should also (a) provide means for workers to communicate their views and ideas, and (b) explain to workers those management actions that on their face are antagonistic to quality.[8]

Communication with customers is sometimes misunderstood as one of the basic strategies used in a total quality setting. It does not mean asking customers what new products should be invented. Customers don't typically think in these terms. In his book *Out of the Crisis*, W. Edwards Deming writes:

> The customer is not in a good position to prescribe a product or service that will help him in the future. The producer is in a far better position than the consumer to invent new design and new service. Would anyone that owned an automobile in 1905 express a desire for pneumatic tires, had you asked him what he needed? Would I, carrying a precise pocketwatch, have suggested a tiny calculator and quartz timepiece?"[9]

All of the market research in the world won't spare the entrepreneur the anxiety of dealing with the inescapable element of risk. However, having taken the risk to produce a product, communicating with customers about that product can ensure that it gets the best possible reception in the marketplace and that it changes as the needs of customers change.

Know Your Customer's Operations

As a supplier to other companies (customers), it is important to know their operations. The more that is known about a customer's operations, the easier it will be to provide products that meet their needs.

What does the customer do with our product? How is it used? Is our product part of a larger assembly? Does the customer use our product in the way we expect or in some different way? Does the customer modify our product in any way? What processes does the customer use in working with our product?

Knowing the answers to questions such as these can help a supplier improve customer satisfaction. The answers to these types of questions can lead to such benefits as the following:

- *Product enhancements.*   By knowing a customer's operations, suppliers might be able to modify their products to better fit in with them. They might also be able to add attributes that will make the product even more attractive to the customer.
- *Improved productivity.*   By knowing a customer's operations, suppliers might be able to propose process modifications that will improve their productivity.
- *Internal improvements.*   By knowing a customer's operations, suppliers might learn facts that lead to internal improvements in quality, productivity, and design in their own organization.

Customers don't always use a product in the way a supplier assumes. By getting to know customers and their operations, suppliers have not just made process and product improvements; some have actually started new product lines. In any case, the better suppliers know their customers' operations, the better they can serve them. The better suppliers serve their customers, the greater the likelihood of satisfied, long-term customers.

CUSTOMER SATISFACTION PROCESS

Customer focus is more than just sending out surveys. Customer focus is part of a process that leads to continual improvements in the organization that, in turn, result in customer satisfaction. Resources are limited. Consequently, they must be applied where they will do the most to improve customer satisfaction and customer retention. The process described in the following list will help meet all these goals:[10]

- Determine who your customers are.
- Determine what attributes of your product/service are most important to your customers.
- Arrange these attributes in the order of importance indicated by your customers.
- Determine your customers' level of satisfaction with each of these attributes.
- Tie results of customer feedback to your processes.
- Develop a set of metrics (measurements) that tell how you are performing and which areas within the process are having the greatest impact on performance.
- Implement measurements at the lowest possible level in the organization.
- Work on those processes that relate to attributes that have high importance, but low customer satisfaction rating.
- Work on those areas within the process that offer the greatest opportunity to improve.
- Update customer input and feedback on a continual basis. Then, as process improvements correspondingly improve customer satisfaction, move on to the next most important process improvements.
- Maintain open continual communication with all stakeholders on what is being done, why, what results are expected, and when.
- Aggregate metrics organization-wide into a format for management review on a continual basis. Adjust as necessary.

CUSTOMER RETENTION

Customer satisfaction is a fundamental cornerstone of total quality. An organization develops a customer focus to be better able to satisfy its customers. Consequently, forward-looking organizations use customer satisfaction data to measure success. But measuring customer satisfaction alone is not enough. Another important measure of success is customer retention.

Certainly customer satisfaction is the critical component in customer retention, but the two factors are not necessarily synonymous. A customer satisfied is not always a customer retained. Frederick F. Reichheld makes the point that although "it may seem intuitive that increasing customer satisfaction will increase retention and therefore profits, the facts are contrary. Between 65 and 85 percent of customers who defect say they were satisfied or very satisfied with their former supplier."[11] Reichheld's findings suggest that there is more to customer retention than just customer satisfaction.

Many business leaders assume that having acquired customers they need only provide high-quality products and services to retain them. Michael W. Lowenstein calls this the *"myth of customer satisfaction."*[12] According to Lowenstein, "Conventional wisdom of business, academia, and the consulting community is that . . . if satisfied, the customer will remain loyal. Reality proves that customer loyalty or retention is a more complex, yet more definitive indicator of quality performance."[13]

It is important to understand what Lowenstein is saying here. Is he saying that customer satisfaction is not important? No, of course not. Customer satisfaction is critical, but it is a means to an end, not an end in itself. The desired end is customer retention. What Lowenstein is saying is that organizations should measure success based on customer retention data rather than on customer satisfaction data. The issue is not whether customers are satisfied with the organization's products or services. It is whether they are satisfied enough to be retained. Satisfied customers will sometimes defect in spite of their satisfaction, if for no other reason than curiosity about a competitor or the ever present lure of variety. How, then, can an organization go beyond just satisfying its customers to retaining them? The short answer to this question is as follows:

> To retain customers over the long term, organizations must turn them into partners and proactively seek their input rather than waiting for and reacting to feedback provided after a problem has occurred.

The following strategies can help organizations go beyond just satisfying customers to retaining them over the long term. These strategies will help organizations operationalize the philosophy of turning customers into partners.

Be Proactive; Get Out in Front of Customer Complaints

Many organizations make the mistake of relying solely on feedback from customers for identifying problems, the most widely used mechanism in this regard being the customer complaint process. Feedback-based processes, although necessary and useful, have three glaring weaknesses. First, they are activated by problems customers have already experienced. Even if these problems are solved quickly, the customer who complains has already had a negative experience with the organization. Such experiences are typically remembered—even if only subconsciously—no matter how well the organization responds.

The second weakness of feedback-oriented processes is that they are based on the often invalid assumption that dissatisfied customers will take the time to lodge a complaint. Some will, but many won't. Some people are just too busy to take the time to complain. Others give their feedback by simply going elsewhere. In a survey of retail customers conducted by the Institute for Corporate Competitiveness (ICC), 67% of respondents said they would simply go elsewhere if dissatisfied rather than take the time to complain.[14] Retail customers don't necessarily have the same characteristics as customers of service or production organizations. However, the ICC survey still points to a fundamental weakness with customer complaint processes that rely on information collected ex post facto.

The third weakness of customer complaint processes is that the information they provide is often too sketchy to yield an accurate picture of the problem. This situation can result in an organization wasting valuable resources chasing after symptoms rather than solving root causes. The weaknesses associated with after-the-fact processes do not mean that organizations should stop collecting customer feedback. On the contrary, customer feedback can be important when used to supplement the data collected using input-based processes.

Customer input is customer information provided *before* a problem occurs. An effective vehicle for collecting customer input is the focus group. *Focus groups* consist of customers who agree to meet periodically with representatives of the organization for the purpose of pointing out issues before they become problems. Focus groups can provide a mechanism for overcoming all three of the weaknesses associated with feedback systems. Participants point out weaknesses or potential issues to the organization's representatives so that they can be dealt with preemptively. Focus group input does not depend on the willingness of customers to lodge complaints. Participants agree to provide input at periodic meetings before becoming members of the group.

A variation on the focus group concept is the input group. The purpose of both types of groups is to provide input the organization can use to improve their processes and products and services. The difference between the two is that focus group participants meet together for group discussion. Input group participants provide their data individually, usually by mail, telephone, or facsimile machine. They do not meet together for group interaction.

The focus group approach can also solve the problem of sketchy information. In a focus group, there is discussion, debate, and give and take. This type of interaction gives the organization's representatives opportunities to dig deeper and deeper until they get beyond symptoms to root causes. Input provided by one participant will often trigger input from another.

To be effective, the focus group must consist of participants who understand what they are being asked to do. The organization is well served by neither sycophants nor witch hunters. What is needed is information that is thoroughly thought out and objective, given in the form of open, honest, constructive criticism. Membership of the focus group should change periodically to bring in new ideas and a broader cross-section of input.

Other methods for collecting customer input include hiring test customers and conducting periodic surveys of a representative sample of the customer base. *Test customers* are individuals who do business with the organization and report their perceptions to designated representatives of the organization. This method can backfire unless employees are fully informed that it is a method the organization uses. This does not mean that employees should know who the test customers are. They shouldn't, or this method will lose its value. However, they should know that any customer they interact with might be a test customer.

Customer surveys conducted periodically can help identify issues that may become problems. If this method is used, the survey instrument should be brief and to the point. One of the surest ways to turn off customers is to ask them to complete a lengthy survey instrument. Some type of reward should be associated with completing the survey that says "Thank you for your valuable time and assistance." Each time a survey is conducted, care should be taken to select a different group of customers. Asking the same people to complete surveys over and over is sure to turn off even the most loyal customers.

Collect Both Registered and Unregistered Complaints

Many organizations make the mistake of acting solely on what customers say in complaints instead of going beyond what is said to include what is not being said. Lowenstein calls this phenomenon the *Iceberg Complaint Model.*[15] In other words, registered complaints from customers are just the tip of the iceberg that is seen above the surface of the water. A much larger portion of the iceberg floats quietly beneath the surface. For this reason it is important for organizations to collect both registered and unregistered complaints.

Focus groups—already discussed—are an excellent way to solicit unregistered complaints. Customer surveys and test customers can also serve this purpose. Another way to get at that part of the iceberg that floats beneath the surface is the *follow-up interview.* With this method, customers who have registered complaints are contacted either in person or by telephone to discuss their complaint in greater depth. This approach gives representatives of the organization the opportunity to ask clarifying questions and to request suggestions.

Another way to get at unregistered complaints is to use the organization's sales representatives as collectors of customer input. Sales representatives are the employees who have the most frequent face-to-face contact with customers. If properly trained concerning what to look for, what to ask, and how to respond, sales personnel can bring back invaluable information from every sales call. In addition to providing sales personnel with the necessary training, organizations should also provide them with appropriate incentives for collecting customer input. Otherwise they may fall into the trap of simply agreeing with the customer about complaints received, thereby undermining the customer relationship even further.

ESTABLISHING A CUSTOMER FOCUS

Companies that have successfully established a customer focus share a number of common characteristics. In his book *The Customer Driven Company*, Richard C. Whitely suggests that these characteristics can be divided into seven clusters:[16]

- *Vision, commitment, and climate.*   A company with these characteristics is totally committed to satisfying customer needs. This commitment shows up in everything the company does. Management demonstrates by deeds and words that the customer is important, that the organization is committed to customer satisfaction, and that customer needs take precedence over internal needs. One way such organizations show their commitment to customers and establish a climate in which customer satisfaction prevails is by making the matter of being customer focused a major factor in all promotions and pay increases.
- *Alignment with customers.*   Customer-driven companies align themselves with their customers. Customers are included when anyone in the organization says "we." Alignment with customers manifests itself in several ways, including the following: customers play a consultative role in selling, customers are never promised more than can be delivered, employees understand what product attributes the customers value most, and customer feedback and input are incorporated into the product development process.
- *Willingness to find and eliminate customers' problems.*   Customer-driven companies work hard to continually identify and eliminate problems for customers. This willingness manifests itself in the following ways: customer complaints are monitored and analyzed; customer feedback is sought continually; and internal processes, procedures, and systems that create no value for customers are identified and eliminated.
- *Use of customer information.*   Customer-driven companies not only collect customer feedback but also use it and communicate it to those who need it to make improvements. The use of customer information manifests itself in the following ways: all employees know how the customer defines quality, employees at different levels are given opportunities to meet with customers, employees know who the "real" customer is, customers are given information that helps them develop realistic expectations, and employees and managers understand what customers want and expect.
- *Reaching out to customers.*   Customer-driven companies reach out to their customers. In a total quality setting, it is never enough to sit back and wait for customers to give evaluative feedback. A competitive global marketplace demands a more assertive approach. Reaching out to customers means doing the following:

  Making it easy for customers to do business
  Encouraging employees to go beyond the normal call of duty to please customers
  Attempting to resolve all customer complaints
  Making it convenient and easy for customers to make their complaints known

- *Competence, capability, and empowerment of people.*   Employees are treated as competent, capable professionals and are empowered to use their judgment in doing what is necessary to satisfy customer needs. This means that all employees have a thorough understanding of the products they provide and the customer's needs relating to those products. It also means that employees are given the resources and support needed to meet the customer's needs.

■ *Continuous improvement of products and processes.* Customer-driven companies do what is necessary to continuously improve their products and the processes that produce them. This approach to doing business manifests itself in the following ways: internal functional groups cooperate to reach shared goals, best practices in the business are studied (and implemented wherever they will result in improvements), research and development cycle time is continually reduced, problems are solved immediately, and investments are made in the development of innovative ideas.

These seven clusters of characteristics can be used as a guide in establishing a customer focus. The first step is a self-analysis in which it is determined which of these characteristics are present in the organization and which are missing. Characteristics that are missing form the basis for an organization-wide implementation effort.

RECOGNIZING THE CUSTOMER-DRIVEN ORGANIZATION

Is a given organization customer driven? In today's competitive business environment, the answer to this question must be yes. Since this is the case, it is important for quality professionals to be able to recognize a customer-driven organization and to be able to articulate its distinguishing characteristics.

According to Whitely, a customer-driven organization can be recognized by the following characteristics:[17]

■ *Reliability*—an organization that dependably delivers what is promised on time every time

■ *Assurance*—an organization that is able to generate and convey trust and confidence

■ *Tangibles*—an organization that pays attention to the details in all aspects of its operations

■ *Empathy*—an organization that conveys a real interest in its customers

■ *Responsiveness*—an organization that is willingly attentive to customer needs

In addition to these characteristics, Thomas Cartin and Donald Jacoby recommend that quality professionals look for the following management factors:[18]

■ *Internal support.* Has the organization developed the right structure management, and employee skills (organizational development)?

■ *Use of knowledge.* Does the organization systematically collect customer feedback and continually assess changing customer expectations and needs?

■ *Use of metrics.* Has the organization established market-based performance measures that are tied to customer needs rather than internal factors?

■ *Communication.* Does the organization regularly communicate progress toward meeting customer needs to its employees?

By applying the characteristics and management factors set forth in this section quality professionals can recognize customer-driven organizations. They can also quickly surmise where problems and shortcomings exist and develop a clearly focused plan of improvement.

ENDNOTES

1. Peter R. Scholtes, *The Team Handbook* (Madison, WI: Joiner Associates, 1992), 2–6.
2. Giorgio Merli, *Total Manufacturing Management* (Cambridge, MA: Productivity Press, 1990), 6–7.

3. Scholtes, 1–11.
4. "Zytec Employees Can Spend up to $1,000 to Satisfy Customers," *Total Quality Newsletter* 3 (July 1992): 4.
5. Scholtes, 5–48.
6. Scholtes, 5-48–5-49.
7. Joseph M. Juran, *Juran on Leadership for Quality* (New York: Free Press, 1989), 101.
8. Juran, 314.
9. W. Edwards Deming, *Out of the Crisis* (Cambridge: Massachusetts Institute of Technology, Center for Advanced Engineering Study, 1991), 167.
10. Michael Butchko, Mike Butchko Consulting, "Customer Satisfaction Process," a white paper dated July 25, 1995. Used with permission.
11. Frederick F. Reichheld, "Loyalty-Based Management," *Harvard Business Review* (March–April 1993), 71.
12. Michael W. Lowenstein, *Customer Retention—An Integrated Process for Keeping Your Best Customers* (Milwaukee: ASQC Quality Press, 1995), 1.
13. Lowenstein, 10.
14. "Customer Complaint Survey," Final Report, Institute for Corporate Competitiveness, 1994.
15. Lowenstein, 37–39.
16. Richard C. Whitely, *The Customer Driven Company: Moving from Talk to Action* (New York: Addison-Wesley, 1991), 221–225.
17. Whitely, 109–111.
18. Thomas J. Cartin and Donald J. Jacoby, *A Review of Managing Quality and a Primer for the Certified Quality Manager Exam* (Milwaukee, WI: ASQC Quality Press, 1997), 106.

Empowerment

Involving people in decisions made relating to their work is a fundamental principle of good management. With total quality, this principle is taken even further. First, employees are involved not only in decision making but also in the creative thought processes that precede decision making. Second, not only are employees involved; they are empowered. This chapter explains the concepts of involvement and empowerment, their relationship, and how they can be used to improve competitiveness.

EMPOWERMENT DEFINED

Employee involvement and empowerment are closely related concepts, but they are not the same. In a total quality setting, employees should be empowered:

> James Monroe, CEO of a midsized electronics manufacturing firm, decided more than a year ago to get his employees involved, as a way to improve work and enhance his company's competitiveness. He called his managers and supervisors together, explained his idea, and had suggestion boxes placed in all departments. At first, the suggestion boxes filled to overflowing. Supervisors emptied them once a week, acted on any suggestions they thought had merit, and discarded the rest. After a couple of months, employee suggestions dwindled down to one or two a month. Worse, recent suggestion forms have contained derisive remarks about the company and its suggestion system. Productivity has not improved, and morale is worse than before. Monroe is at a loss over what to do. Employee involvement was supposed to help, not hurt.

In this example, the CEO involved his employees, but he failed to empower them. To understand the difference, it is necessary to begin with an understanding of employee involvement as a concept. Peter Grazier, Author of *Before It's Too Late: Employee*

Involvement . . . An Idea Whose Time Has Come, describes employee involvement as follows:

> What is Employee Involvement? It's a way of engaging employees at all levels in the thinking processes of an organization. It's the recognition that many decisions made in an organization can be made better by soliciting the input of those who may be affected by the decision. It's an understanding that people at all levels of an organization possess unique talents, skills, and creativity that can be of significant value if allowed to be expressed.[1]

What, then, is empowerment? Stated simply, *empowerment* is employee involvement that matters. It's the difference between just having input and having input that is heard, seriously considered, and followed up on whether it is accepted or not.

Most employee involvement systems fail within the first year, regardless of whether they consist of suggestion systems, regularly scheduled brainstorming sessions, daily quality circles, one-on-one discussions between an employee and a supervisor, or any combination of the various involvement methods. The reason is simple: they involve but do not empower employees. Employees soon catch on to the difference between having input and having input that matters. Without empowerment, involvement is just another management tool that doesn't work. This is what went wrong in the example of the electronics manufacturing firm at the beginning of this section. The CEO implemented a system that involved but did not empower his employees.

Management Tool or Cultural Change

The management strategies developed over the years to improve productivity, quality, cost, service, and response time would make a long list. Is empowerment just another of these management tools, another strategy to add to the list? This is an important question, and it should be dealt with in the earliest stages of implementing empowerment.

Employees who have been around long enough to see several management innovations come and go may be reluctant to accept empowerment if they see it as just another short-lived management strategy. This is known as the WOHCAO syndrome (pronounced WO-KAY-O). WOHCAO is short for "Watch out, here comes another one."

Successful implementation of empowerment requires change in the corporate culture—a major new direction in how managers think and work. The division of labor between managers and workers changes with empowerment. Grazier describes the cultural change required by empowerment as follows:

> It's not something "nice" we do for our employees to make them feel better. It's an understanding that it's everyone's obligation—part of the job—to constantly look for better ways of doing things. It's part of the job to ask questions and raise issues of concern, to get them out on the table so they can be resolved. How else can we get better? Above all, employee involvement is not simply another management tool, but a major change of direction in the way we lead our workforce. It's a change that affects the culture of the workplace as we've come to know it in this century, and, therefore, it's a change that must be implemented with great care and attention.[2]

Empowerment Does Not Mean Abdication

It is not uncommon for traditional managers to view empowerment as an abdication of power. Such managers see involvement and empowerment as turning over control of the company to the employees. In reality, this is hardly the case. Empowerment involves actively soliciting input from those closest to the work and giving careful thought to that input.

Pooling the collective minds of all people involved in a process, if done properly, will enhance rather than diminish managers' power. It increases the likelihood that the information on which decision makers base their decisions is comprehensive and accurate. Managers do not abdicate their responsibility by adopting empowerment. Rather, they increase the likelihood of making the best possible decision and, thereby, more effectively carry out their responsibility.

RATIONALE FOR EMPOWERMENT

Traditionally, working hard was seen as the surest way to succeed. With the advent of global competition and the simultaneous advent of automation, the key to success became not just working hard but also working smart. In many cases, decision makers in business and industry interpreted working smart as adopting high-tech systems and automated processes. These smarter technologies have made a difference in many cases. However, better technology is just one aspect of working smarter, and it's a part that can be quickly neutralized when the competition adopts a similar or even better technology.

An aspect of working smart that is often missing in the modern workplace is involving and empowering employees in ways that take advantage of their creativity and promote independent thinking and initiative on their part. In other words, what's missing is empowerment. Creative thinking and initiative from as many employees as possible will increase the likelihood of better ideas, better decisions, better quality, better productivity, and, therefore, better competitiveness. The rationale for empowerment is that it represents the best way to bring the creativity and initiative of the best employees to bear on improving the company's competitiveness.

Empowerment and Motivation

According to Dr. Isobel Pfeiffer and Dr. Jane Dunlop of the University of Georgia:

> Empowerment is the key to motivation and productivity. An employee who feels he or she is valued and can contribute is ready to help and grow in the job. Empowerment enables a person to develop personally and professionally so that his or her contributions in the workplace are maximized.[3]

Empowerment is sometimes seen by experienced managers as just another name for participatory management. However, there is an important distinction between the two. Participatory management is about managers and supervisors asking for their employees' help. Empowerment is about getting employees to help themselves, each other, and the company. This is why empowerment can be so effective in helping maintain a high level of motivation among employees. It helps employees develop a sense of ownership of their jobs and in the company. This, in turn, leads to a greater willingness on the part of employees to make decisions, take risks in an effort to make improvements, and speak out when they disagree.

INHIBITORS OF EMPOWERMENT

The primary inhibitor of empowerment, resistance to change, is an indigenous characteristic of human nature. According to Grazier:

> Resistance to employee involvement is real. Since 1970, when the concept of employee involvement was first introduced in America, resistance has caused the failure of many employee involvement efforts. It is puzzling that a concept that produces both tangible benefits for the organization and a great deal of satisfaction for the workforce creates so much resistance. Resistance to empowerment, when it occurs, comes from three different groups: employees, unions, and management. Although the greatest resistance to empowerment traditionally comes from management, sometimes consciously and sometimes unconsciously, resistance from unions and employees should not be overlooked.[4]

Resistance from Employees and Unions

Earlier in this chapter, the WOHCAO syndrome was discussed. In this syndrome, employees have experienced enough flash-in-the-pan management strategies that either did not work out or were not followed through on that they have become skeptical.

In addition to skepticism, there is the problem of inertia. Resistance to change is natural. Even positive change can be uncomfortable for employees because it involves new and unfamiliar territory. However, when recognized for what they are, skepticism and inertia can be overcome. Strategies for doing so are discussed later in this chapter.

Unions can be another source of resistance when implementing empowerment. Because of the traditional adversarial relationship between organized labor and management, unions may be suspicious of management's motives in implementing empowerment. They might also resent an idea not originated by their own organization. However, unions' greatest concern is likely to be how empowerment will affect their future. If union leaders think it will diminish the need for their organization, they will throw up roadblocks.

Resistance from Management

Even if employees and labor unions support empowerment, it will not work unless management makes a full and whole-hearted commitment to it. Kizilos says that "too many companies are attempting only half-heartedly to empower employees."[5] What he means is that companies are attempting to implement empowerment without first making the necessary fundamental changes in organizational structure or management style.

The importance of management commitment cannot be overemphasized. Employees take their cues from management concerning what is important, what the company is committed to, how to behave, and all other aspects of the job.

Grazier summarizes the reasons behind management resistance to empowerment as follows: insecurity, personal values, ego, management training, personality characteristics of managers, and exclusion of managers.[6]

Workforce Readiness

An inhibitor of employee empowerment that receives little attention in the literature is workforce readiness. Empowerment will fail quickly if employees are not ready to be empowered. In fact, empowering employees who are not prepared for the responsibilities involved can be worse than not empowering them at all. On the other hand, a lack of readiness—even though it may exist—should not be used as an excuse for failing to empower employees. The challenge to management is twofold: (a) determine whether the workforce is ready for empowerment; and (b) if it is not ready, get it ready.

How, then, does one know whether the workforce is ready for empowerment? One rule of thumb is that the more highly educated the workforce, the more ready they will be for empowerment. Because well-educated people are accustomed to critical thinking, they are experienced in decision making, and they tend to make a point of being well informed concerning issues that affect their work. This does not mean, however, that less educated employees should be excluded. Rather, it means that they may need to be made ready before being included.

In determining whether employees are ready for empowerment, ask the following questions:

■ Are the employees accustomed to *critical thinking*?
■ Are the employees knowledgeable of the decision-making process and their role with regard to it?
■ Are employees fully informed of the *big picture* and where they fit into it?

Unless the answer to all three of these questions is yes, the workforce is not ready for empowerment. An empowered employee must be able to think critically. It should be second nature for such an employee to ask such questions as the following: Is there a better way to do this? Why do we do it this way? Could the goal be accomplished some other way? Is there another way to look at this problem? Is this problem really an opportunity to improve things?

These are the types of questions that lead to continuous improvement of processes and more effective solutions to problems. These are the sorts of questions that empowered employees should ask all the time about everything. Employees who are unaccustomed to asking questions such as these should be taught to do so before being empowered.

Employees should understand the decision-making process both on a conceptual level and on a practical level (e.g., how decisions are made in their organization). Being empowered does not mean making decisions. Rather, it means being made a part of the decision-making process. Before empowering employees, it is important to show them what empowerment will mean on a practical level. How will they be empowered? Where do they fit into the decision-making process? They also need to know the boundaries. What decisions are they able to make themselves or within their work team? Employees should know the answers to all of these questions before being empowered.

Organizational Structure and Management Practices

Most of the resistance to empowerment is attitudinal, as the inhibitors explained so far show. However, a company's organizational structure and its management practices can also mitigate against the successful implementation of empowerment. Before attempting to implement empowerment, the following questions should be asked:

■ How many layers of management are there between workers and decision makers?
■ Does the employee performance appraisal system encourage or discourage initiative and risk taking?
■ Do management practices encourage employees to speak out against policies and procedures that inhibit quality and productivity?

Employees, like most people, will become frustrated if their suggestions have to work their way through a bureaucratic maze before reaching a decision maker. Prompt feedback on suggestions for improvement is essential to the success of empowerment. Too many layers of managers who can say no between employees and decision makers who can say yes will inhibit and eventually kill risk taking and initiative on the part of employees.

Risk-taking employees will occasionally make mistakes or try ideas that don't work. If this reflects negatively in their performance appraisals, initiative will be replaced by a play-it-safe approach. This also applies to constructive criticism of company policies and management practices. Are employees who offer constructive criticism considered problem solvers or troublemakers? Managers' attitudes toward constructive criticism will determine whether they receive any. A positive, open attitude in such cases is essential. The free flow of constructive criticism is a fundamental element of empowerment.

MANAGEMENT'S ROLE IN EMPOWERMENT

Management's role in empowerment can be stated simply. It is to do everything necessary to ensure successful implementation and ongoing application of the concept. The three words that best describe management's role in empowerment are *commitment, leadership*, and *facilitation*. All three functions are required to break down the barriers and overcome the inherent resistance often associated with implementation of empowerment or with any other major change in the corporate culture.

Grazier describes the manager's role in empowerment as demonstrating the following support behaviors:

- Exhibiting a supportive attitude
- Being a role model
- Being a trainer
- Being a facilitator
- Practicing management by walking around (MBWA)
- Taking quick action on recommendations
- Recognizing the accomplishments of employees[7]

IMPLEMENTING EMPOWERMENT

Creating a workplace environment that is positive toward and supportive of empowerment so that risk taking and individual initiative are encouraged is critical. Targeting and overcoming inhibitors of empowerment is also critical. It is also critical when implementing empowerment to consider the employee's point of view and to put some helpful empowerment vehicles in place.

Considering the Employee's Point of View

Employee risk taking and initiative require a certain environment and a certain set of conditions. Richard Hamlin describes this phenomenon from the employee's perspective by saying, "As an employee, I am willing to learn new thinking and doing skills if the following conditions exist: 1) I see that I will be better off for the learning, and 2) I perceive that a non-punitive pathway is available to me."[8] Hamlin goes on to describe what managers can do to facilitate risk taking and initiative on the part of employees as follows:[9]

- Believe in their ability to be successful.
- Be patient and give them time to learn.
- Provide direction and structure.
- Teach them new skills in small, incremental steps.
- Ask questions that challenge them to think in new ways.
- Share information with them sometimes to just build rapport.
- Give timely, understandable feedback and encourage them throughout the learning process.
- Offer alternative ways of performing tasks.
- Exhibit a sense of humor and demonstrate care for them.
- Focus on results and acknowledge personal improvement.

Having managers who are able to view the workplace from the point of view of employees can help build their trust, and trust is critical to empowerment. Trust shows employees that they are believed in.

Putting Vehicles in Place

A number of different types of vehicles can be used for soliciting employee input and channeling it to decision makers. Such vehicles range from simply walking around the workplace and asking employees for their input, to periodic brainstorming sessions, to regularly scheduled quality circles. Widely used methods that are typically the most effective are explained in the following subsections.

Brainstorming

With brainstorming, managers serve as catalysts in drawing out group members. Participants are encouraged to share any idea that comes to mind. All ideas are considered valid. Participants are not allowed to make judgmental comments or to evaluate the suggestions made. Typically one member of the group is asked to serve as a recorder. All ideas suggested are recorded, preferably on a marker board, flip chart, or another medium that allows group members to review them continuously.

After all ideas have been recorded, the evaluation process begins. Participants are asked to go through the list one item at a time, weighing the relative merits of each. This process is repeated until the group narrows the choices to a specified number. For example, managers may ask the group to reduce the number of alternatives to three, reserving the selection of the best of the three to themselves.

Brainstorming can be an effective vehicle for collecting employee input and feedback, particularly if managers understand the weaknesses associated with it and how they can be overcome. Managers interested in soliciting employee input through brainstorming should be familiar with the concepts of groupthink and groupshift. These two concepts can undermine the effectiveness of brainstorming and other group techniques.

Groupthink is the phenomenon that exists when people in a group focus more on reaching a decision than on making a good decision.[10] A number of factors can contribute to groupthink, including the following: overly prescriptive group leadership, peer pressure for conformity, group isolation, and unskilled application of group decision-making techniques. Mel Schnake recommends the following strategies for overcoming groupthink:[11]

- Encourage criticism.
- Encourage the development of several alternatives. Do not allow the group to rush to a hasty decision.
- Assign a member or members to play the role of devil's advocate.
- Include people who are not familiar with the issue.
- Hold last-chance meetings. When a decision is reached, arrange a follow-up meeting a few days later. After group members have had time to think things over, they may have second thoughts. Last-chance meetings give employees an opportunity to voice their second thoughts.

Groupshift is the phenomenon that exists when group members exaggerate their initial position hoping that the eventual decision will be what they really want.[12] If group members get together prior to a meeting and decide to take an overly risky or overly conservative view, it can be difficult to overcome. Managers can help minimize the effects of groupshift by discouraging reinforcement of initial points of view and by assigning group members to serve as devil's advocates.

Nominal Group Technique

The nominal group technique (NGT) is a sophisticated form of brainstorming involving five steps. In the first step, the manager states the problem and clarifies if necessary to make sure all group members understand. In the second step, each group member silently records his or her ideas. At this point, there is no discussion among group members. This strategy promotes free and open thinking unimpeded by judgmental comments or peer pressure.

In the third step, the ideas of individual members are made public by asking each member to share one idea with the group. The ideas are recorded on a marker board or flip chart. The process is repeated until all ideas have been recorded. Each idea is numbered. There is no discussion among group members during this step. Taking the ideas one at a time from group members ensures a mix of recorded ideas, making it more difficult for members to remember what ideas belong to which member.

In the fourth step, recorded ideas are clarified to ensure that group members understand what is meant by each. A group member may be asked to explain an idea, but no comments or judgmental gestures are allowed from other members. The member clarifying the ideas is not allowed to make justifications. The goal in this step is simply to ensure that all ideas are clearly understood.

In the final step, the ideas are voted on silently. There are a number of ways to accomplish this. One simple technique is to ask all group members to record the numbers of their five favorite ideas on five separate 3×5 cards. Each member then prioritizes his or her five cards by assigning them a number ranging from one (worst idea) to five (best idea). The cards are collected and the points assigned to ideas are recorded on the marker board or flip chart. After this process has been accomplished for all five cards of all group members, the points are tallied. The idea receiving the most points is selected as the best idea.

Quality Circles

A quality circle is a group of employees that meets regularly for the purpose of identifying, recommending, and making workplace improvements. A key difference between quality circles and brainstorming is that quality circle members are volunteers who convene themselves and conduct their own meetings. Brainstorming sessions are typically convened and conducted by a manager. A quality circle has a team leader who acts as a facilitator, and the group may use brainstorming, NGT, or other group techniques; however, the team leader is typically not a manager and may, in fact, be a different group member at each meeting. Quality circles meet regularly before, during, or after a shift to discuss their work, anticipate problems, propose workplace improvements, set goals, and make plans.

Suggestion Boxes

This vehicle is perhaps the oldest method used for collecting employee input and feedback. It consists of placing receptacles in convenient locations into which employees may put written suggestions.[13] Figure 5–1 is an example of a form used for making written suggestions that would be put in a suggestion box. By examining this form, you can see that suggestions made at Manufacturing Technologies Corporation are logged in, acknowledged, and tracked. They may be made by individuals or teams, and they require an explanation of the current situation, proposed improvements, and benefits expected from the improvements.

Walking and Talking

Simply walking around the workplace and talking with employees can be an effective way to solicit input. This approach is sometimes referred to as management by walking around (MBWA). An effective way to prompt employee input is to ask questions. This approach may be necessary to get the ball rolling, particularly when empowerment is still new and not yet fully accepted by employees. In such cases, it is important to ask the right questions. According to Joseph T. Straub, managers should "ask open-ended unbiased questions that respect . . . workers' views and draw them out."[14]

MANAGEMENT'S ROLE IN SUGGESTION SYSTEMS

Management's role in the implementation and operation of suggestion systems can be divided into several steps, as shown in Figure 5–2.

Establishing Policy   This step involves developing the policies that will guide the suggestion system. Such policies should clearly describe the company's commitment to the suggestion system, the types of rewards that will be used, how suggestions will be evaluated, and how the suggestion system itself will be evaluated.

Manufacturing Technologies Corporation
Two Industrial Park
Fort Walton Beach, Florida 32549

Name _____
        (Individual or team making the suggestion)

Date Submitted _____

Department _____

Telephone Extension _____

Suggestion (Explain current situation, proposed change, and expected benefits.)

_____

_____

_____

_____

_____

Date Received _____

Date Logged In _____

Date Suggestion Was Acknowledged _____

Current Status _____

_____

_____

(Attach sketches or other illustrative material.)

Figure 5–1
Suggestion Form

Setting Up the System   This step involves putting the system in place to do the following:

1. Solicit and collect employee input.
2. Acknowledge and log in suggestions.
3. Monitor suggestions.
4. Implement or reject suggestions.

Promoting the Suggestion System   This step involves generating interest and participation in the suggestion system on the part of employees. The following strategies are effective in promoting suggestion systems:

■ Sharing the company's policies concerning suggestion systems in frank and open group meetings that encourage questions and discussion
■ Sponsoring suggestion competitions
■ Asking employees for their input concerning how to increase participation

Evaluating Suggestions and the Suggestion System   This step involves teaching supervisors and managers how to evaluate individual suggestions and the overall system. Both of these topics are explained later in this chapter.

Figure 5–2
Management's Role in Suggestion Systems

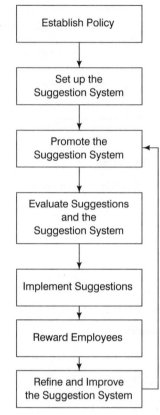

Implementing Suggestions   This is a critical step. If good suggestions are not promptly implemented, the system will lose credibility no matter how well the other components work.

Rewarding Employees   Rewards for suggestions that are implemented can take many forms. Cash awards, public recognition, and paid vacations are a few of the more widely used types of rewards. In addition to rewarding individuals, it is important to reward teams and/or departments that make the most accepted suggestions or that save the largest total dollar amount for the company over a specified period of time.

Refining and Improving the Suggestion System   As with any system, weaknesses will show up in the day-to-day operation of the suggestion system. It is important to identify and correct these weaknesses. Continual improvements to the system in such key areas as soliciting input, tracking and monitoring suggestions, evaluating suggestions, and decreasing the time required to implement suggestions are important. Employee input should be solicited concerning refinements to the suggestion system itself, just as it is with the processes and systems used to do company's work.

IMPROVING SUGGESTION SYSTEMS

Many U.S. companies now accept employee input as not only desirable but necessary to compete. The growing realization of the importance of employee input is encouraging. However, if only 32% of the suggestions received in a given year are adopted, a clear need exists to improve both individual suggestions and suggestion systems.

Improving Suggestion Processing

The *suggestion system* is the collection of processes used to solicit, collect, evaluate, and adopt or turn down suggestions. According to Bob Scharz, author of *The Suggestion System: A Total Quality Process*, a good suggestion system meets all of the following criteria:[15]

- All suggestions receive a formal response.
- All suggestions are responded to immediately.
- Performance of each department in generating and responding to suggestions is monitored by management.
- System costs and savings are reported.
- Recognition and awards are handled promptly.
- Good ideas are implemented.
- Personality conflicts are minimized.

From these criteria, it can be seen that operating a suggestion system involves more than having employees toss ideas into a box, accepting some, and throwing the rest away. The best suggestion systems require that ideas be submitted in writing on a special form (see again Figure 5–1). Such forms make it easier for employees to submit suggestions and for employers to give immediate and formal responses. The forms ease the job of logging in suggestions and tracking them through the entire life of the idea until it is either adopted or rejected.

An approach that is being used increasingly is the computer-aided suggestion system. With such a system, employees make suggestions in writing on a form such as the one in Figure 5–1. Immediate acknowledgment is provided by the manager of the department in question or by the employee who administers the computer-aided system. Acknowledged suggestions are entered into a personal computer and monitored from that point forward, creating a database of suggestions that can be updated continually. As the status of a suggestion changes, the information is fed to the employee who operates the computer-aided system and the database is updated.

Such a system makes it easy for management to assess the performance of individual departments in generating and responding to suggestions and to monitor system costs and savings. Figure 5–3 is an example of a monthly report prepared from the database of a computer-aided suggestion system. In this example, only suggestions that receive a rating of 5 or higher on a scale of 1–10 are included. This is a wise approach on the part of the company because it will deter employees from submitting frivolous suggestions in an attempt to show higher numbers on the monthly report.

Improving Individual Suggestions

To make good suggestions, employees need to know how to do two things:

- Identify problems and formulate ideas for improvement.
- Clearly and concisely communicate their ideas in written and graphic form.

Teaching employees to use the following three steps in formulating ideas for workplace improvements can increase the number of ideas they generate and the quality of those ideas:

1. *Problem identification* involves identifying situations that differ from the desired result. At this point, no attempt is made to determine the cause of the problem, nor should a cause be too quickly assumed. Employees should be helped to learn to approach problems systematically and in a step-by-step manner. In this step,

Monthly Report
Suggestions Submitted by Department*
GLOBAL MANUFACTURING, INC.
Crestview Plant

Department	Number of Suggestions for Month of __March__
	(scale: 5, 10, 15, 20)
Administration	▓▓▓▓▓▓▓▓▓▓▓▓▓▓▓▓▓▓
Sales/Marketing	▓▓▓▓▓▓▓▓▓▓▓
Purchasing	▓▓▓▓▓▓▓▓▓▓▓▓▓▓
Engineering	▓▓▓▓▓▓▓▓▓▓▓▓▓
Electronics Mfg.	▓▓▓▓▓▓▓
Metal Fabrication	▓▓▓▓▓▓▓▓
Wire Harness Fab.	▓▓▓▓▓▓
Quality Assurance	▓▓▓▓▓▓▓▓▓▓▓▓▓▓
Shipping/Receiving	▓▓▓
Accounting	▓▓▓▓▓▓▓▓▓▓▓▓

* Only suggestions receiving a rating of 5 (out of a possible 10) or higher are included in this report.

Figure 5–3
Suggestion Summary Report

the task is to record problem situations that are candidates for improvements. This is the "What is wrong?" step.

2. *Research* comes next. Whereas the preceding step (problem identification) deals with the issue of what is wrong, research deals with why it is wrong. Research into why is important because, if done properly, it can prevent the expenditure of resources treating symptoms rather than causes.

    To help determine causes, numerous total quality tools can be used, including Pareto charts, cause-and-effect diagrams, bar graphs, check sheets, scatter diagrams, and histograms. These tools are discussed later in this book.

3. *Idea development* involves the development of ideas for solving the problem identified in the first step by eliminating the cause identified in the second step. Figure 5–4 is a checklist that can be used as an aid in the development of ideas. This checklist corresponds with the JHRA's recommendation that the following factors be considered when developing ideas: subject matter, purpose, location, sequence, people, method, and cost.[16]

When ideas for improvement have been formulated, employees should write them down using graphics (sketches, diagrams, graphs, charts and/or photos) as appropriate to illustrate their ideas. One of the essential steps in improving individual suggestions is to help employees learn to write clear, concise, definitive descriptions of their ideas for improvement.

Managers can help employees improve their individual suggestions by coaching them to do the following:

■ Briefly, succinctly, and clearly explain the current situation that creates the problem.

■ Get to the heart of the proposed change with no preliminaries or rationalizations, and be specific.

Broad Question	Examples of More Specific Questions
What?	• What is the task in question?
	• What would happen if the task were eliminated?
Why?	• Why is the task performed?
	• Why haven't alterations been attempted?
Where?	• Where is the task performed?
	• Where else might the task be performed?
When?	• When is the task performed?
	• When else could the task be performed?
Who?	• Who performs the task?
	• Who else might perform the task?
How?	• How is the task performed?
	• How else could the task be performed?
How much?	• How much does the task cost now?
	• How much will it cost after improvements are implemented?

Figure 5–4
Idea Development Checklist

■ Prepare illustrations to clarify the proposed change, in every case where this is appropriate.

■ Explain the expected benefits using quantifiable terms (percentages, dollars, time, numbers, or whatever is appropriate).

■ Take nothing for granted. Assume that the decision makers who will read the suggestion know nothing about the situation.

EVALUATING SUGGESTIONS

Perhaps the most critical and most difficult aspect of operating an effective suggestion system is evaluating suggestions. How does one recognize a good idea? How can one respond to a poor idea without discouraging the employee who submitted it? These questions are dealt with in this section.

Evaluating and Rating Suggestions

Employees are often in the best position to make suggestions for workplace improvements, but in the final analysis it is still the manager who must decide whether the proposed idea is feasible. It's a lot like a loan officer in a bank trying to decide which loan applicants will pay off and which will be deadbeats. There is an art and a science to the process, and those who succeed are the ones who are best able to select the winners most consistently. This also applies to managers who evaluate employee suggestions.

The Japan Human Relations Association recommends the following points be remembered when evaluating employee suggestions:[17]

■ Not all suggestions can be evaluated fairly while sitting at your desk. Questions about a suggestion should be resolved by going to the source and discussing them with the suggester. Be patient and find out what the suggester is trying to say. If necessary, help the suggester improve his or her suggestion.

■ A suggestion that looks bad or outdated on the surface may still contain good ideas. Study all suggestions carefully. Don't jump to conclusions and discount the whole suggestion if part of it is bad.

■ Be generous with first-time suggesters to build up their confidence. If possible, try to accept the suggestion, or at least part of it, making revisions to the suggestion if necessary.

■ With employees who are experienced in suggestion making, evaluate their suggestions carefully and challenge them to set higher goals.

■ Consider the level of the suggester when evaluating a suggestion. If you underestimate a person's ability, he or she will not be challenged and grow. If you expect too much, you will discourage the worker's creativity and initiative.

■ Suggesters will be anxious for feedback. Evaluate suggestions promptly. If a delay is unavoidable, notify the suggester of the reason for it.

■ If possible, notify the suggester of evaluation results in person. Make sure to add a few words of encouragement, especially if the idea is not accepted. When providing a written notice, always add a positive comment along with the result.

■ A suggestion is often the product of the suggester and his or her supervisor. Remember to compliment all contributors for their efforts.

Establish a formal rating system for evaluating suggestions to ensure consistency. The rating system should provide a means for quantifying the results of evaluations and have at least the following components:

■ *Criteria and criteria rating.*   Criteria are the factors considered most important in assessing the feasibility of employee suggestions. The Evaluation and Rating Form in Figure 5–5 contains three such criteria. They are the expected benefit of the sug-

Poultry Processing, Inc. Highway 90 East DeFuniak Springs, FL 32816			Suggestion Identification No. _____	
Rating Criteria	Criteria Rating (0–10)	Weight Factor	Weighted Score	Comments
Expected benefit of suggestion		5.0		
Time until benefits will be realized		3.0		
Successful suggestions submitted in past		2.0		
TOTAL SCORE				
Criteria Rating X Weight Factor = Weighted Criteria Score				
Conversion Scale (For Total Score)				

Category 1 . . . . 10 points   Category 6 . . . . 60 points
Category 2 . . . . 20 points   Category 7 . . . . 70 points
Category 3 . . . . 30 points   Category 8 . . . . 80 points
Category 4 . . . . 40 points   Category 9 . . . . 90 points
Category 5 . . . . 50 points   Category 10 . . . . 100 points

Figure 5–5
Employee Suggestion Evaluation and Rating Form

gestion, time that will elapse before benefits will begin to be realized, and how successful suggestions made by the individual or team submitting this suggestion have been in the past (track record). The actual criteria might differ from company to company. However, within a company they should be the same for all departments, to ensure consistency and fairness. Each criterion must be assigned a numerical score or rating. This is the key step in the entire process. It requires judgment, common sense, a thorough knowledge of the situation in question, and an open mind. The example in Figure 5–5 uses a scale of 0 to 10 for each criterion.

- *Weight factors and weighted scores.* Weight factors accommodate the fact that some criteria are more important than others. The individual rating for each criterion is multiplied times its assigned weight factor to determine its weighted score. In Figure 5–5, the expected benefit criterion is assigned a weight factor of 5.0; the time criterion, 3.0; and the final criterion, 2.0. A criterion rating of 7.0 multiplied by a weight factor of 5.0 results in a weighted score of 35.0 points.

- *Total points and conversion scale.* The weighted scores are added together to determine the total score for the suggestion. Just as numerical scores are converted to letter grades in a college class, the total score for the suggestion is converted to a level, category, or grade of suggestion. The conversion scale in Figure 5–5 converts numerical scores to categories. Company policy will typically set the minimum level or category of suggestion that will be considered beyond the supervisory level. For example, Poultry Processing, Inc., might establish a company policy that only suggestions categorized as Level 6 or higher (see Figure 5–5) will be considered for implementation.

HANDLING POOR SUGGESTIONS

Even the best suggestion system will not completely eliminate bad suggestions. In such cases, suggestions must be rejected. "Never adopt a bad idea because you feel sorry for someone or feel you 'owe' him a break,"[18] advises George Milite in an article with the poignant title "When an Employee's Idea Is Just Plain Awful." However, when rejecting a suggestion, precautions should be taken to maintain the employee's interest and morale.

Milite recommends the following strategies for rejecting poor suggestions in a positive manner:[19]

- *Listen carefully.* Give employees an opportunity to explain their suggestions in greater detail. Aspects of the suggestion that have merit or could be developed to make the suggestion more viable may not have been completely explained on the suggestion form.

- *Express appreciation.* Be sure to let employees know that their suggestions are appreciated. Encourage future suggestions. The message to leave them with is this: "This suggestion didn't work out, but your effort is appreciated and your ideas are valued. I am looking forward to your next suggestion."

- *Carefully explain your position.* Don't make excuses or blame the company, higher management, or anyone else. Explain the reasons the suggestion is not feasible, and do so in a way that will help employees make more viable suggestions in the future.

- *Encourage feedback.* Encourage feedback from employees. You may have overlooked an aspect to the suggestion, or the suggester may not fully understand the reasons for the rejection. Solicit sufficient feedback to ensure that you understand the employees and they understand you.

- *Look for compromise.* It may be possible to use all or a portion of a suggestion if it is modified. Never adopt a bad idea, but if a suggestion can be modified to make even part of it worthwhile, a compromise solution may be possible.

ACHIEVING FULL PARTICIPATION

To achieve optimum potential, a suggestion program must have a broad base of participating employees. The closer a company comes to full participation, the better its chances are of having a successful suggestion system. When an effective suggestion system is in place and working, managers can focus on the next level of concern: achieving full participation. This goal amounts to removing hidden barriers, encouraging new employees to get involved, and coaching reluctant employees.

Removing Hidden Barriers

Hidden barriers are the less obvious factors that inhibit participation. They can vary widely from company to company and can be difficult to detect. However, the hidden factors that are most often present are these:

- *Negative behavior.* Attitudes and behaviors of personnel to whom suggestions are submitted can inhibit participation. This point applies regardless of whether the recipient is a supervisor or a clerk at a central suggestion submittal point. Negative facial expressions, harsh voice tones, or nonsupportive comments by such persons can turn away tentative employees.

- *Poor writing skills.* Employees with poor writing skills are not uncommon in an age when illiteracy has become commonplace. Because attempts at writing suggestions will call attention to the problem, employees with poor writing skills are not likely to participate in the suggestion program. This problem can be overcome through coaching, counseling, and training.

- *Fear of rejection.* Most human beings fear rejection and, as a result, avoid situations that might subject them to it. Consequently, some employees will not participate in the suggestion program for fear of having their suggestions rejected. This problem can be avoided by working with such employees to help them formulate their initial suggestions and build confidence, which is the key to overcoming fear of rejection.

- *Inconvenience.* Employees are reluctant to participate in suggestion programs that are inconvenient. Are suggestion forms complicated? Is the central submittal point too far away or available only at limited times? Do suggestions need the approval of several people before they can be submitted? Inconveniences such as these will discourage participation.

These are just four examples of hidden barriers that can work against full participation. Talking with employees, observing, and listening can help identify others. Perhaps the most effective way to discover hidden barriers is to use the suggestion program to improve the suggestion program. Encourage employees to submit their ideas concerning factors that discourage participation. Allowing anonymous suggestions in such cases is a good strategy for encouraging honest constructive criticism of the system.

Encouraging New Employees

New employees may experience a natural reluctance to participate in the suggestion program. Encouragement, support, and coaching can overcome this reluctance. The JHRA recommends the following measures for dealing with reluctant new employees:[20]

- Teach new employees why suggestions are important (i.e., quality improvements, efficiency, cost reduction, productivity improvements, better safety, etc.).

- Make a small improvement in the new employee's job, write it up as an example of a suggestion, and use the example to show how the process works.

- Have the new employee work cooperatively with an experienced employee to develop one or more suggestions.

■ Give positive feedback on the new employee's first several tries. Be careful to point out ways the suggestions might be improved.

Coaching Reluctant Employees

Coaching is an important ingredient in all aspects of total quality. In fact, with total quality the middle manager's role is more like that of the coach than that traditionally associated with managers. We typically associate coaching with developing, encouraging, training, and monitoring the performance of athletes. These tasks also apply in the modern workplace where managers concentrate on developing, encouraging, training, and monitoring employees who are members of teams.

Coaching reluctant employees in an attempt to promote participation is similar to coaching athletes in that it involves applying both the carrot and the stick, or, said another way, it entails both encouraging and challenging employees. However, before doing either, it is important to identify the cause of the reluctance.

If an employee wants to make improvements but will not make suggestions, hidden barriers to his or her participation must exist. If the employee doesn't care, he or she may not be in the right job. Occasionally an employee's personal goals will be at odds with those of the work team or the company. If an employee has the requisite knowledge and skills needed to make a contribution but isn't, he or she may be playing the wrong position on the work team. This is a common occurrence on athletic teams: a seemingly talented athlete is not performing well in his or her current position but when moved to another position suddenly comes to life and starts helping the team. A similar result can sometimes be achieved in the workplace.

In such cases, managers must play the role of career coach. Career coaching is a way of getting the right employees into the right jobs and having them take responsibility for their performance, for participating in continual improvement efforts, and for contributing to the work team's performance.

G. M. Sturman recommends the following five-point approach to career coaching that managers can use to better match employees and jobs and to transform employees from noncontributors to self-starters who participate actively in workplace improvement efforts:[21]

■ *Assess.*   Help employees form a clear picture of their interests and aptitudes as well as their strengths and weaknesses.

■ *Investigate.*   Help employees investigate all possible opportunities within the company. Would they be better suited to another department, another team, or even another job?

■ *Match.*   Having assessed interests, aptitudes, strengths, and weaknesses and investigated opportunities, help employees find the optimum match available to them in the company.

■ *Choose.*   Work with employees to help them make the choice that is best for them.

■ *Manage.*   Help employees develop a personal career management plan that will result in the accomplishment of their goals. Do they need additional education and training to get the job they really want? Would another job within the company more fully meet their needs? Their plan should be a road map that guides reluctant employees from where they are to where they can make a contribution to the company while simultaneously meeting their personal goals.

ENDNOTES

1.   Peter B. Grazier, *Before It's Too Late: Employee Involvement . . . An Idea Whose Time Has Come* (Chadds Ford, PA: Teambuilding, 1989), 8.

2.   Grazier, 14.

3. Isobel Pfeiffer and Jane Dunlop, "Increasing Productivity through Empowerment," *Supervisory Management* (January 1990): 11–12.
4. Grazier, 90–91.
5. Peter Kizilos, "Crazy about Empowerment?" *Training* (December 1990): 56.
6. Grazier, 87.
7. Grazier, 129–142.
8. Richard Hamlin, "A Practical Guide to Empowering Your Employees," *Supervisory Management* (April 1991): 8.
9. Hamlin, 8.
10. D. G. Myers and H. Lamm, "The Group Polarization Phenomenon," *Psychological Bulletin* 85 (1976): 602–627.
11. Mel E. Schnake, *Human Relations* (New York: Macmillan, 1990), 285–286.
12. R. D. Clark, "Group-Induced Shift toward Risk: A Critical Appraisal," *Psychological Bulletin* 80 (1971): 251–270.
13. J. R. Jablonski, *Implementing TQM* (London: Pfeiffer, 1992), 98.
14. J. T. Straub, "Ask Questions First to Solve the Right Problems," *Supervisory Management* (October 1991): 7.
15. B. Scharz, *The Suggestion System: A Total Quality Process* (Cambridge, MA: Productivity Press, 1991), 92.
16. Japan Human Relations Association, ed., *The Idea Book* (Cambridge, MA: Productivity Press, 1988), 33.
17. Japan Human Relations Association, 140.
18. George Milite, "When an Employee's Idea Is Just Plain Awful," *Supervisory Management* (October 1991): 3.
19. Milite, 3.
20. Japan Human Relations Association, 108.
21. G. M. Sturman, "The Supervisor as a Career Coach," *Supervisory Management* (November 1990): 6.

CHAPTER SIX

Leadership and Change

Leadership is an intangible concept that produces tangible results. It is referred to sometimes as an art and at other times as a science. In reality, leadership is both an art and a science.

The impact of good leadership can be readily seen in any organization where it exists. Well-led organizations, whether they are large companies or small departments within a company, share several easily identifiable characteristics:

- High levels of productivity
- Positive, can-do attitudes
- Commitment to accomplishing organizational goals
- Effective, efficient use of resources
- High levels of quality

LEADERSHIP DEFINED

Leadership can be defined in many different ways, partly because it has been examined from the perspective of so many different fields of endeavor. Leadership has been defined as it applies to the military, athletics, education, business, industry, and many other fields. For the purpose of this book, *leadership* is defined as it relates specifically to total quality:

> Leadership is the ability to inspire people to make a total, willing, and voluntary commitment to accomplishing or exceeding organizational goals.

This definition contains a key concept that makes it particularly applicable in a total quality setting: the concept of inspiring people. Inspiring people is a higher order of human interaction than motivating them, which is a concept more frequently used in defining leadership. *Inspiration*, as used here, means motivation that has been internalized and therefore comes from within employees, as opposed to motivation that is simply a temporary response to external stimuli. Motivated employees commit to the organization's goals. Inspired employees make those goals their own. When employees are inspired, the total, willing, and voluntary commitment described in the definition follows naturally.

What Is a Good Leader?

Good leaders come in all shapes, sizes, genders, ages, races, political persuasions, and national origins. They do not look alike, talk alike, or even work alike. However, good leaders do share several common characteristics. These are the characteristics necessary to inspire people to make a total, willing, and voluntary commitment. Regardless of their backgrounds, good leaders exhibit the following characteristics:

- Balanced commitment
- Positive influence
- Good communication skills
- Positive role model
- Persuasiveness

Good leaders are committed to both the job to be done and the people who must do it, and they are able to strike the appropriate balance between the two. Good leaders project a positive example at all times. They are good role models. Managers who project a "Do as I say, not as I do" attitude will not be effective leaders. To inspire employees, managers must be willing to do what they expect of workers, do it better, do it right, and do so consistently. If, for example, dependability is important, managers must set a consistent example of dependability. If punctuality is important, a manager must set a consistent example of punctuality. To be a good leader, a manager must set a consistent example of all characteristics that are important on the job.

Good leaders are good communicators. They are willing, patient, skilled listeners. They are also able to communicate their ideas clearly, succinctly, and in a nonthreatening manner. They use their communication skills to establish and nurture rapport with employees. Good leaders have influence with employees and use it in a positive manner. *Influence* is the art of using power to move people toward a certain end or point of view. The power of managers derives from the authority that goes with their jobs and the credibility they establish by being good leaders. Power is useless unless it is converted to influence. Power that is properly, appropriately, and effectively applied becomes positive influence.

Finally, good leaders are persuasive. Managers who expect people to simply do what they are ordered to do will have limited success. Those who are able to use their communication skills and influence to persuade people to their point of view and to help people make a total, willing, and voluntary commitment to that point of view can have unlimited success.

LEADERSHIP FOR QUALITY

Leadership for quality is leadership from the perspective of total quality. It is about applying the principles of leadership set forth in the preceding section in such a way as to continually improve work methods and processes. Leadership for quality is based on the philosophy that continually improving work methods and processes will, in turn, improve quality, cost, productivity, and return on investment.

This is the philosophy articulated in the Deming Chain Reaction developed by quality pioneer W. Edwards Deming. Deming's philosophy is that each improvement in work methods and processes initiates a chain reaction that results in the following:[1]

- Improved quality
- Decreased costs
- Improved productivity
- Decreased prices
- Increased market share
- Longevity in business
- More jobs
- Improved return on investment

Principles of Leadership for Quality

The principles of leadership for quality parallel those of total quality. In his book *The Team Handbook,* Peter R. Scholtes summarizes these principles as described in the following sections.[2]

Customer Focus

Leadership for quality requires a customer focus. This means the organization's primary goal is to meet or exceed customer expectations in a way that gives the customer lasting value. In a total quality setting, there are both internal and external customers. Internal customers are other employees within the organization whose work depends on the work that precedes theirs. External customers are people who purchase and/or use the organization's products.

Obsession with Quality

Obsession with quality is an attitude that must be instilled and continually nurtured by leaders in an organization. It means that every employee aggressively pursues quality in an attempt to exceed the expectations of customers, internal and external.

Recognizing the Structure of Work

Leadership for quality requires that work processes be analyzed to determine their appropriate structural makeup (organization, order of steps, tools used, motion required, etc.). When the optimum structure is in place, work processes should be analyzed, evaluated, and studied continually in an attempt to improve them.

Freedom through Control

Control in a total quality setting refers to human control of work methods and processes. All too often in the age of high technology, the "tail wags the dog" in that machines run people instead of people running machines. Leaders must ensure that managers and employees take control of work processes and methods by working together to standardize them. The goal is to reduce variations in output by eliminating variations in how work is done.

Unity of Purpose

One of the most important responsibilities of a leader is to articulate the organization's mission clearly and accurately so that all employees understand it, believe in it, and commit to it. When there is unity of purpose, all employees pull together toward the same end.

Looking for Faults in Systems

Quality pioneers W. Edwards Deming and Joseph M. Juran believed that 85% of an organization's failures are failures of management-controlled systems. In their opinion, employees who do the work can control only 15% of what causes failures. Leadership for quality requires a change in focus from assessing blame for problems to assessing systems in an attempt to ferret out and correct systemic problems.

Teamwork

Rugged individualism has long been a fundamental element of the American character. The strong, silent stranger who rides into town and single-handedly runs out the bad guys (the character typified by Clint Eastwood over the years) has always had popular appeal in the United States. Individual performance has been encouraged and rewarded in the American workplace since the Industrial Revolution. Not until competition among companies became global in nature did it become necessary to apply a principle that has been known for years—that a team of people working together toward a common goal can outperform a group of individuals working toward their own ends. Leadership for quality requires team building and teamwork. These critical topics are covered later in this chapter.

Continuing Education and Training

In the age of high technology, the most important machine in the workplace is the human mind. This is the premise of the book *Thinking for a Living* by Ray Marshall, who serves as professor of economics at the University of Texas and is co-chair of the Commission on the Skills of the American Workforce.[3] Continued learning at all levels is a fundamental element of total quality. Working hard no longer guarantees success. In the age of high technology, it is necessary to work hard and work smart.

LEADERSHIP STYLES

Leadership styles have to do with how people interact with those they seek to lead. Leadership styles go by many different names. However, most styles fall into the following categories:

- Autocratic
- Democratic
- Participative
- Goal-oriented
- Situational

Autocratic Leadership

Autocratic leadership is also called *directive* or *dictatorial leadership*. People who take this approach make decisions without consulting the employees who will have to implement them or who will be affected by them. They tell others what to do and expect them to comply obediently. Critics of this approach say that although it can work in the short run or in isolated instances, in the long run it is not effective. Autocratic leadership is not appropriate in a total quality setting.

Democratic Leadership

Democratic leadership is also called *consultive* or *consensus leadership*. People who take this approach involve the employees who will have to implement decisions in making them. The leader actually makes the final decision, but only after receiving the input and recommendations of team members. Critics of this approach say the most popular decision is not always the best decision and that democratic leadership, by its nature, can result in the making of popular decisions as opposed to right decisions. This style can also lead to compromises that ultimately fail to produce the desired result.

Participative Leadership

Participative leadership is also known as *open, free-reign,* or *nondirective leadership.* People who take this approach exert little control over the decision-making process. Rather, they provide information about the problem and allow team members to develop strategies and solutions. The leader's job is to move the team toward consensus. The underlying assumption of this style is that workers will more readily accept responsibility for solutions, goals, and strategies that they are empowered to help develop. Critics of this approach say consensus building is time-consuming and works only if all people involved are committed to the best interests of the organization.

Goal-Oriented Leadership

Goal-oriented leadership is also called *results-based* or *objective-based leadership.* People who take this approach ask team members to focus only on the goals at hand. Only strategies that make a definite and measurable contribution to accomplishing organizational goals are discussed. The influence of personalities and other factors unrelated to the specific goals of the organization is minimized. Critics of this approach say it can break down when team members focus so intently on specific goals that they overlook opportunities and/or potential problems that fall outside of their narrow focus. Advocates of total quality say that results-oriented leadership is too narrowly focused and often centered on the wrong concerns.

Situational Leadership

Situational leadership is also called *fluid* or *contingency leadership.* People who take this approach select the style that seems to be appropriate based on the circumstances that exist at a given time. In identifying these circumstances, leaders consider the following factors:

- Relationship of the manager and team members
- How precisely actions taken must comply with specific guidelines
- Amount of authority the leader actually has with team members

Depending on what is learned when these factors are considered, the manager decides whether to take the autocratic, democratic, participative, or goal-oriented approach. Under different circumstances, the same manager would apply a different leadership style. Advocates of total quality reject situational leadership as an attempt to apply an approach based on short-term concerns instead of focusing on the solution of long-term problems.

LEADERSHIP STYLE IN A TOTAL QUALITY SETTING

The appropriate leadership style in a total quality setting might be called participative leadership taken to a higher level. Whereas participative leadership in the traditional sense involves soliciting employee input, in a total quality setting it involves soliciting input from empowered employees, listening to that input, and acting on it. The key difference between the traditional participative leadership and participative leadership from a total quality perspective is that, with the latter, employees providing input are empowered.

Collecting employee input is not new. However, collecting input, logging it in, tracking it, acting on it in an appropriate manner, working with employees to improve weak suggestions rather than simply rejecting them, and rewarding employees for improvements that result from their input—all of which are normal in a total quality setting—extend beyond the traditional approach to participative leadership.

BUILDING AND MAINTAINING A FOLLOWING

Managers can be good leaders only if the people they hope to lead will follow them willingly and steadfastly. Followership must be built and, having been built, maintained. This section is devoted to a discussion of how managers can build and maintain followership among the people they hope to lead.

Popularity and the Leader

Leadership and popularity are not the same thing. However, many managers confuse popularity with leadership and, in turn, followership. An important point to understand in leading people is the difference between popularity and respect. Long-term followership grows out of respect, not popularity. Good leaders *may* be popular, but they *must* be respected. Not all good leaders are popular, but they are all respected.

Managers occasionally have to make unpopular decisions. This is a fact of life for leaders, and it is why leadership positions are sometimes described as lonely positions. Making an unpopular decision does not necessarily cause a leader to lose followership, provided the leader is seen as having solicited a broad base of input and given serious, objective, and impartial consideration to that input. Correspondingly, leaders who make inappropriate decisions that are popular in the short run may actually lose followership in the long run. If the long-term consequences of a decision turn out to be detrimental to the team, team members will hold the leader responsible, particularly if the decision was made without first collecting and considering employee input.

Leadership Characteristics That Build and Maintain Followership

Leaders build and maintain followership by earning the respect of those they lead. Some characteristics of leaders that build respect are explained in the following paragraphs:

- *Sense of purpose.*   Successful leaders have a strong sense of purpose. They know who they are, where they fit in the overall organization, and the contributions their areas of responsibility make to the success of the organization.
- *Self-discipline.*   Successful leaders develop discipline and use it to set an example. Through self-discipline, leaders avoid negative self-indulgence, inappropriate displays of emotion such as anger, and counterproductive responses to the everyday pressures of the job. Through self-discipline, leaders set an example of handling problems and pressures with equilibrium and a positive attitude.
- *Honesty.*   Successful leaders are trusted by their followers. This is because they are open, honest, and forthright with other members of the organization and with themselves. They can be depended on to make difficult decisions in unpleasant situations with steadfastness and consistency.
- *Credibility.*   Successful leaders have credibility. Credibility is established by being knowledgeable, consistent, fair, and impartial in all human interaction; by setting a positive example; and by adhering to the same standards of performance and behavior expected of others.
- *Common sense.*   Successful leaders have common sense. They know what is important in a given situation and what is not. They know that applying tact is important when dealing with people. They know when to be flexible and when to be firm.
- *Stamina.*   Successful leaders must have stamina. Frequently they need to be the first to arrive and the last to leave. Their hours are likely to be longer and the pressures they face more intense than those of others. Energy, stamina, and good health are important to those who lead.

- *Commitment.*   Successful leaders are committed to the goals of the organization, the people they work with, and their own ongoing personal and professional development. They are willing to do everything within the limits of the law, professional ethics, and company policy to help their team succeed.
- *Steadfastness.*   Successful leaders are steadfast and resolute. People do not follow a person they perceive to be wishy-washy and noncommittal. Nor do they follow a person whose resolve they question. Successful leaders must have the steadfastness to stay the course even when it becomes difficult.

Pitfalls That Can Undermine Followership

The previous section explained several positive characteristics that will help managers build and maintain the respect and, in turn, followership of those they hope to lead. Managers should also be aware of several common pitfalls that can undermine followership and the respect managers must work so hard to earn:

- *Trying to be a buddy.*   Positive relations and good rapport are important, but leaders are not the buddies of those they lead. The nature of the relationship does not allow it.
- *Having an intimate relationship with an employee.*   This practice is both unwise and unethical. A positive manager–employee relationship cannot exist under such circumstances. Few people can succeed at being the lover and the boss, and few things can damage the morale of a team so quickly and completely.
- *Trying to keep things the same when supervising former peers.*   The supervisor–employee relationship, no matter how positive, is different from the peer–peer relationship. This can be a difficult fact to accept and a difficult adjustment to make. But it is an adjustment that must be made if the peer-turned-supervisor is going to succeed as a leader.

LEADERSHIP VERSUS MANAGEMENT

Leadership and management, although both are needed in the modern workplace, are not the same thing. To be a good leader and a good manager, one must know the difference between the two concepts. According to John P. Kotter, leadership and management "are two distinctive and complementary systems of action."[4] Kotter lists the following differences between management and leadership:[5]

- Management is about coping with complexity; leadership is about coping with change.
- Management is about planning and budgeting for complexity; leadership is about setting the direction for change through the creation of a vision.
- Management develops the capacity to carry out plans through organizing and staffing; leadership aligns people to work toward the vision.
- Management ensures the accomplishment of plans through controlling and problem solving; leadership motivates and inspires people to want to accomplish the plan.

Warren Bennis quotes Field Marshall Sir William Slim, who led the British Army's brilliant reconquest of Burma during World War II, on drawing the distinction between management and leadership:

> Managers are necessary; leaders are essential. . . . Leadership is of the Spirit, compounded of personality and vision. . . . Management is of the mind, more a matter of accurate calculation, statistics, methods, timetables, and routine.[6]

Trust Building and Leadership

Trust is a necessary ingredient for success in the intensely competitive modern workplace. It means, in the words of D. Zielinski and C. Busse, "employees who can make hard decisions, access key information, and take initiative without fear of recrimination from management, and managers who believe their people can make the right decisions."[7]

Building trust requires leadership on the part of managers. Trust-building strategies include the following:

- *Taking the blame but sharing the credit.*   Managers who point the finger of blame at their employees, even when the employees are at fault, do not build trust. Leaders must be willing to accept responsibility for the performance of people they hope to lead. Correspondingly, when credit is due, leaders must be prepared to spread it around appropriately. Such unselfishness on the part of managers builds trust among employees.

- *Pitching in and helping.*   Managers can show leadership and build trust by rolling up their sleeves and helping when a deadline is approaching. A willingness to get their hands dirty when circumstances warrant it helps managers build trust among employees.

- *Being consistent.*   People trust consistency. It lets them know what to expect. Even when employees disagree with managers, they appreciate consistent behavior.

- *Being equitable.*   Managers cannot play favorites and hope to build trust. Employees want to know that they are treated not just well, but as well as all other employees. Fair and equitable treatment of all employees will help build trust.

LEADERSHIP AND ETHICS

In his book *Managing for the Future: The 1990's and Beyond*, Drucker discusses the role of ethics in leadership.[8] Setting high standards of ethical behavior is an essential task of leaders in a total quality setting. Drucker summarizes his views regarding ethical leadership in the modern workplace as follows:

> What executives do, what they believe and value, what they reward and whom, are watched, seen, and minutely interpreted throughout the whole organization. And nothing is noticed more quickly—and considered more significant—than a discrepancy between what executives preach and what they expect their associates to practice. The Japanese recognize that there are really only two demands of leadership. One is to accept that rank does not confer privileges; it entails responsibilities. The other is to acknowledge that leaders in an organization need to impose on themselves that congruence between deeds and words, between behavior and professed beliefs and values, that we call personal integrity.[9]

Drucker speaks to ethical behavior on the part of corporate executives who hope to lead their companies effectively. The concept also applies to managers at all levels and any other person who hopes to lead others in a total quality setting.

LEADERSHIP AND CHANGE

In a competitive and rapidly changing marketplace, businesses are constantly involved in the development of strategies for keeping up, staying ahead, and/or setting new directions. What can managers do to play a positive role in the process? David Shanks recommends the following strategies:[10]

- Have a clear vision and corresponding goals.
- Exhibit a strong sense of responsibility.
- Be an effective communicator.
- Have a high energy level.
- Have the will to change.

Shanks developed these strategies to help executives guide their companies through periods of corporate stress and change, but they also apply to other personnel. These characteristics of good leaders apply to any manager at any level who must help his or her organization deal with the uncertainty caused by change.

Facilitating Change as a Leadership Function

The following quote by consultant Donna Deeprose carries a particularly relevant message for today's competitive environment:

> In an age of rapidly accelerating technology, restructuring, repositioning, downsizing, and corporate takeovers, change may be the only constant. Is there anything you can do about it? Of course there is. You can make change happen, let it happen to you, or you can stand by while it goes on around you.[11]

Deeprose divides people into three categories, based on how they handle change: driver, rider, or spoiler.[12] People who are drivers lead their organizations in new directions as a response to change. People who just go along reacting to change as it happens rather than getting in front of it are riders. People who actively resist change are spoilers. Deeprose gives examples of how a driver would behave in a variety of situations:[13]

- In viewing the change taking place in an organization, drivers stay mentally prepared to take advantage of the change.
- When facing change about which they have misgivings, drivers step back and examine their own motivations.
- When a higher manager has an idea that has been tried before and failed, drivers let the boss know what problems were experienced earlier and offer suggestions for avoiding the problems this time.
- When a company announces major changes in direction, drivers find out all they can about the new plans, communicate what they learn with their employees, and solicit input to determine how to make a contribution to the achievement of the company's new goals.
- When permission to implement a change that will affect other departments is received, drivers go to these other departments and explain the change in their terms, solicit their input, and involve them in the implementation process.
- When demand for their unit's work declines, drivers solicit input from users and employees as to what changes and new products or services might be needed and include the input in a plan for updating and changing direction.
- When an employee suggests a good idea for change, drivers support the change by justifying it to higher management and using their influence to obtain resources for it while countering opposition to it.
- When their unit is assigned a new, unfamiliar responsibility, drivers delegate the new responsibilities to their employees and make sure that they get the support and training needed to succeed.

These examples show that a driver is a person who exhibits the leadership characteristics necessary to play a positive, facilitating role in helping workers and organizations successfully adapt to change on a continual basis.

HOW TO LEAD CHANGE

Kotter opens the first chapter of his book *Leading Change* with the following statement:

> By any objective measure, the amount of significant, often traumatic, change in organizations has grown tremendously over the past two decades. Although some people predict that most of the reengineering, restrategizing, mergers, downsizing, quality efforts, and cultural renewal projects will soon disappear, I think that is highly unlikely. Powerful macroeconomic forces are at work here, and these forces may grow even stronger over the next few decades. As a result, more and more organizations will be pushed to reduce costs, improve the quality of products and services, locate new opportunities for growth, and increase productivity.[14]

A critical aspect of leadership in today's globally oriented organization involves leading change. To survive and thrive in a competitive environment, organizations must be able to anticipate and respond to change effectively. However, successful organizations don't just respond to change—they get out in front of it.

Change is a constant in today's global business environment. Consequently, organizations must structure themselves for change. In other words, organizations must *institutionalize* the process of change. The model in Figure 6–1 is an effective tool for doing so. The following sections explain the various steps in this model.

Reality of Continual Change

Change is not something organizations do because they want to or because they get bored with the status quo. Rather, it is something they do because they have to. Every organization that must compete in an open marketplace is forced by macroeconomic conditions to constantly reduce costs, improve quality, enhance product attributes, increase productivity, and identify new markets. None of these things can be done without changing the way things are currently done. Employees at all levels in an organization need to understand this point.

If it is true that change is the only constant on the radar screen of the modern organization, why do so many miss the point? A typical reaction is that human beings don't like change. But in reality this is not the case. Employees who object to change typically object to how it is handled, not the fact that it's happening. Kotter identifies the following reasons that employees may not understand the reality of and need for change:[15]

- Absence of a major crisis
- Low overall performance standards
- No view of the big picture
- Internal evaluation measures that focus on the wrong benchmarks
- Insufficient external feedback
- A "kill the messenger" mentality among managers
- Overfocus among employees on the day-to-day stresses of the job
- Too much "happy talk" from executive managers

An organization's senior management team is responsible for eliminating any of these factors except, of course, a major crisis caused by external forces. Creating an artificial crisis is a questionable strategy, but correcting the remainder of these factors is not just appropriate but also advisable.

Performance standards should be based on what it takes to succeed in the global marketplace. Every organization should have a strategic plan that puts the *big picture* into writing, and every employee should know the big picture. Internal evaluation measures should mirror overall performance standards in that they should ensure globally competitive performance.

Figure 6–1
Change Facilitation Model
This model is adapted from John P. Kotter, *Leading Change* (Boston: Harvard Business School Press, 1996), 21.

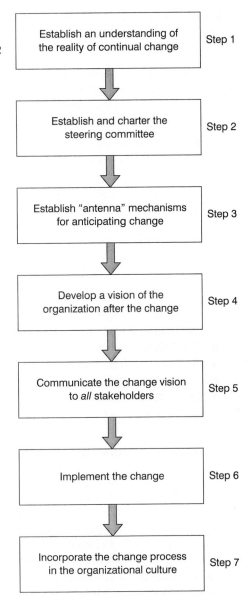

Establish an understanding of the reality of continual change	Step 1
Establish and charter the steering committee	Step 2
Establish "antenna" mechanisms for anticipating change	Step 3
Develop a vision of the organization after the change	Step 4
Communicate the change vision to *all* stakeholders	Step 5
Implement the change	Step 6
Incorporate the change process in the organizational culture	Step 7

Organizations cannot succeed in a competitive environment without systematically collecting, analyzing, and using external feedback. This is how an organization knows what is going on. External feedback is the most effective way to identify the need for change.

Every employee in an organization is a potential for change. Employees attend conferences, read professional journals, participate in seminars, browse on the Internet, and talk to colleagues. If managers welcome feedback from employees, they can turn them collectively, into an effective mechanism for anticipating change. On the other hand, managers who kill the messenger will quickly extinguish this invaluable source of continual feedback.

As a rule, employees will focus most of their attention on their day-to-day duties, which is how it should be. Consequently, management must make a special effort to communicate with employees about market trends and other big picture issues. All such communications with employees should be accurate, thorough, and honest. Managers must make sure they do not go overboard and create panic, but to sugarcoat important issues with happy talk is the same as deceiving employees.

Establish and Charter the Steering Committee

> Major transformations are often associated with one highly visible leader. Consider Chrysler's comeback from near bankruptcy . . . and we think of Lee Iacocca. Mention Wal-Mart's ascension from small-fry to industry leader, and Sam Walton comes to mind. Read about IBM's efforts to renew itself, and the story centers on Lou Gerstner. After a while, one might easily conclude that the kind of leadership that is so critical to any change can come only from a single larger-than-life person.[16]

Although a visible and visionary person can certainly be a catalyst in organizational change, the reality is that one person does not change an organization.

The media like to create the image of the knight on a charging steed who single-handedly saves the company. This story makes good press, but it rarely squares with reality. Organizations that do the best job of handling change have what Kotter calls a "guiding coalition." The *guiding coalition* is a team of people who are committed to the change in question and who can make it happen. Every member of the team should have the following characteristics:

- ■ *Authority.*   Members should have the authority necessary to make decisions and commit resources.
- ■ *Expertise.*   Members should have expertise that is pertinent in terms of the subject change so that informed decisions can be made.
- ■ *Credibility.*   Members must be well respected by all stakeholders so that they will be listened to and taken seriously.
- ■ *Leadership.*   Members should have the leadership qualities necessary to drive the effort. These qualities include those listed here plus influence, vision, commitment, perseverance, and persuasiveness.

Establish Antenna Mechanisms

Leading change is about getting out in front of it. It's about driving change rather than letting it drive you. To do this, organizations must have mechanisms for sensing trends that will generate future change. These "antenna" mechanisms can take many forms, and the more, the better. Reading professional journals, attending conferences, studying global markets, and even reading the newspaper can help identify trends that might affect an organization. So can attentive marketing representatives.

For example, a computer company that markets primarily to colleges and universities learned that two large institutions had adopted a plan to stop purchasing personal computers. Instead, they plan to require all students to purchase their own laptops. The potential for a major change in the organization's business was identified by a marketing representative in the course of a routine call on these universities.

This information allowed the computer company quickly to develop and implement a plan for getting out in front of what is likely to be a nationwide trend. The company now markets laptop computers directly to students through the bookstores at both universities and is marketing the idea to colleges and universities nationwide.

Every employee in an organization should have his or her antennae tuned to the world outside and bring anything of interest to the attention of the steering committee.

Develop a Vision

What will we look like after the change in question has been made? The organization's vision answers this question. Kotter calls a vision a "sensible and appealing picture of the future."[17] The following scenario illustrates how having a cogent vision for change can help employees buy into the change more readily.

Two tour groups are taking a trip together. Each group has its own tour bus and group leader. The route to the destination and stops along the way have been meticulously planned for maximum tourist value, interest, and enjoyment. All members of both groups have agreed to the itinerary and are looking forward to every stop. Unfortunately, several miles before the first scheduled attraction, the tour group leaders receive word that a chemical spill on the main highway will require a detour that will, in turn, require changes to the itinerary.

The tour group leader for group A simply acknowledges the message and tells the bus driver to take the alternate route. To the members of group A he says only that an unexpected detour has forced a change of plans. With no more information than this to go on, the members of group A are confused and quickly become unhappy.

The tour leader for group B, however, is a different sort. She acknowledges the message and then asks the driver to pull over. She then says to the members of group B:

> Folks, we've had a change of plans. A chemical spill on the highway up ahead has closed down the route on which most of today's attractions are located. Fortunately, this area is full of wonderful attractions. Why don't we just have lunch on me at a rustic country restaurant? It's up the road about 5 miles. While you folks are enjoying lunch, I'll hand out a list of new attractions I know you'll want to see. We aren't going to let a little detour spoil our fun!

Because they could see how things would look after the change, the members of group B accepted it and were satisfied. However, the members of group A, because they were not fully informed, became increasingly frustrated and angry. As a result, they went along with the changes only reluctantly and, in several cases, begrudgingly. Group A's tour leader kept his clients in the dark. Group B's leader gave hers a vision.

The six characteristics of an effective vision are as follows:

- *Imaginable.* It conveys a picture that others can see of how things will be after the change.
- *Desirable.* A vision that points to a better tomorrow for all stakeholders will be well received even by those who resist change.
- *Feasible.* To be feasible, a vision must be realistic and attainable.
- *Flexible.* An effective vision is stated in terms that are general enough to allow for initiative in responding to ever-changing conditions.
- *Communicable.* A good vision can be explained to an outsider who has no knowledge of the business.

Communicate the Vision to All Stakeholders

People will buy into the vision only if they know about it. The vision must be communicated to all stakeholders. A good communication package will have at least the following characteristics:

- Simplicity
- Repetitiveness
- Multiple formats
- Feedback mechanisms

The simpler the message, the better. Regardless of the communication formats chosen, keep the message simple and get right to it. Don't beat around the bush, lead with rationalizations, or attempt any type of linguistic subterfuge—what politicians and journalists call "spin."

Repetition is critical when communicating a new message. Messages are like flies. If a fly buzzes past your face just once and moves on, you will probably take no notice of it. But if it keeps coming back persistently, buzzing, and refuses to go away, before long you will take notice of it. Repetition forces employees to take notice.

It is also important to use multiple formats, such as small-group meetings, large-group meetings, newsletter articles, fliers, bulletin board notices, videotaped messages,

E-mail, and a variety of other methods. A combination of visual, reading, and listening vehicles will typically be the most effective.

Regardless of the communication formats used, one or more feedback mechanisms must be in place. In face-to-face meetings, the feedback can be spontaneous and direct. This is always the best form of feedback. However, telephone, facsimile, E-mail, and written feedback can also be valuable.

Implement the Change

Implementing the change is a step that is usually composed of numerous substeps. It involves doing everything necessary to get the change made, which might include some or all of the following substeps:

- Removing structural inhibitors to change
- Enabling employees through training
- Confronting managers and supervisors who continue to resist change
- Planning and generating short-term wins to get the ball rolling
- Eliminating unnecessary interdependencies among functional components of the organization

Incorporate the Change Process

Once an organization has gone through the transformational process of change, both the change itself and the process of change should be incorporated as part of the organization's culture. In other words, two things need to happen. First, the major change that has just occurred must be anchored in the culture so that it becomes the normal way of doing business. Second, the process explained in this section for facilitating change must be institutionalized.

Anchoring the new change in the organizational culture is critical. If this does not occur, the organization will quickly begin to backslide and retrench. The following strategies will help an organization anchor a major change in its culture:

- *Showcase the results.*   In the first place, a change must work to be accepted. The projected benefits of making the change should be showcased as soon as they are realized and, of course, the sooner, the better.
- *Communicate constantly.*   Don't assume that stakeholders will automatically see, understand, and appreciate the results gained by making the change. Talk about results and their corresponding benefits constantly.
- *Remove resistant employees.*   If key personnel are still fighting the change after it has been made and is producing the desired results, give them the "get with it or get out" option. This approach might seem harsh, but employees at all levels are paid to move an organization forward, not to hold it back.

Institutionalizing the process of change is an important and final element of this step. Change is not something that happens once and then goes away. It is a constant in the lives of every person, in every organization. Consequently, the change facilitation model explained here (Figure 6–1) must become part of normal business operation. Antenna mechanisms continue to anticipate change all the time, forever. They feed what they see into the model, and the organization works its way through each step explained here. The better an organization becomes at doing this, the more successful it will be at competing in the global marketplace.

ENDNOTES

1. Peter Scholtes, *The Team Handbook* (Madison, WI: Joiner Associates, 1992), 1–9.
2. Scholtes, 1-11–1-13.

3. Ray Marshall and Marc Tucker, *Thinking for a Living* (New York: Basic Books, 1992).

4. John P. Kotter, "What Leaders Really Do," *Harvard Business Review* 2(3): 103–104.

5. Kotter, 105.

6. W. Bennis, "Leadership in the 21st Century," *Training* (May 1990): 44.

7. D. Zielinski and C. Busse, "Quality Efforts Flourish When Trust Replaces Fear and Doubt," *Total Quality* (December 1990): 103.

8. Peter F. Drucker. *Managing for the Future: The 1990s and Beyond* (New York: Truman Talley Books/Dutton, 1992), 113–117.

9. Drucker, 116–117.

10. D. C. Shanks, "The Role of Leadership in Strategy Development," *Journal of Business Strategy* (January/February 1989): 36.

11. D. Deeprose, "Change: Are You a Driver, a Rider, or a Spoiler?" *Supervisory Management* (February 1990): 3.

12. Deeprose, 3.

13. Deeprose, 3.

14. John P. Kotter, *Leading Change* (Boston: Harvard Business School Press, 1996), 3.

15. Kotter, 40.

16. Kotter, 51.

17. Kotter, 71.

Team Building

OVERVIEW OF TEAM BUILDING

Teamwork is a fundamental element of total quality. The reason for this is simple and practical, as can be seen in the following quote:

> Someone may be great at his or her job, maybe even the best there ever was. But what counts at work is the organization's success, not personal success. After all, if your organization fails, it does not matter how great you were; you are just as unemployed as everyone else.[1]

What Is a Team?

A team is a group of people with a common, collective goal. The collective goal aspect of teams is critical. This point is evident in the performance of athletic teams. For example, a basketball team in which one player hogs the ball, plays the role of the prima donna, and pursues his or her own personal goals (a personal high point total, MVP status, publicity, or something else) will rarely win against a team of players, all of whom pull together toward the collective goal of winning the game.

Rationale for Teams

A team's ability is more than the sum of the abilities of individual members. This is one of the primary reasons for advocating teamwork. Perry L. Johnson describes the rationale for teams as follows:[2]

- Two or more heads are better than one.
- The whole (the team) is greater than the sum of its parts (individual members).
- People in teams get to know each other, build trust, and as a result, want to help each other.
- Teamwork promotes better communication.

It is well established that teams can outperform individuals, provided they are properly handled. A team is not just a group of people. A group of people becomes a team when the following conditions exist:

- Agreement exists as to the team's mission. For a group to be a team and a team to work effectively, all members must understand and agree on the mission.

- Members adhere to team ground rules. A team must have ground rules that establish the framework within which the team's mission is pursued. A group becomes a team when there is agreement as to mission and adherence to ground rules.

- Fair distribution of responsibility and authority exists. Teams do not eliminate structure and authority. Football teams have quarterbacks, and baseball teams have captains. However, teams work best when responsibility and authority are shared and team members are treated as equals.

- People adapt to change. Change is not just inevitable in a total quality setting—it is desirable. Unfortunately, people typically resist change. People in teams should help each other adapt to change in a positive way.

Learning to Work Together

A group of people does not make a team. People in a group do not automatically or magically find ways to work together. Concerning the pressures that can keep a group from becoming a team, teamwork expert Peter R. Scholtes says:

> Dealing with internal group needs that arise from these pressures is as important as the group's external task of making improvements. Yet even teams that grasp the importance of improving quality often underestimate the need for developing themselves as teams. . . . A team that fails to build relationships among its members will waste time on struggle for control and endless discussions that lead nowhere.[3]

One of the reasons teams don't always work as well as they might is certain built-in human factors that, unless understood and dealt with, can work against success. Scholtes describes these factors as follows:[4]

- *Personal identity of team members.*   It is natural for people to wonder where they fit into any organization. This tendency applies regardless of whether the organization is a company or a team within a company. People worry about being an outsider, getting along with other team members, having a voice, and developing mutual trust among team members. The work of the team cannot proceed effectively until team members feel as if they fit in.

- *Relationships among team members.*   Before people in a group can work together, they have to get to know each other and form relationships. When people know each other and care about each other, they will go to great lengths to support each other. Time spent helping team members get acquainted and establish common ground among themselves is time invested well. This point is especially important now that the modern workforce has become so diverse that common ground among team members can no longer be assumed.

- *Identity within the organization.*   This factor has two aspects. The first has to do with how the team fits into the organization. Is its mission a high priority in the company? Does the team have support at the highest management levels? The second aspect of this factor has to do with how membership on a given team will affect relationships with nonteam members. This concern is especially important in the case of task forces and project teams on which team members will want to maintain relationships they have already established with fellow employees who are not on the team. They may be concerned that membership on the team might have a negative impact on their relationships with fellow workers who aren't on the team.

How to Be a Team Leader

Managers in the modern workplace will be called on to lead teams. Mary Walsh Massop recommends the following strategies for being an effective team leader:[5]

- *Be clear on the team's mission.*   The team's first organizational meeting should be used to draft a mission statement. This task, although guided by the team leader, should involve all team members. The statement should explain the reasons for the team's existence and define the limits of its authority. The mission statement will be the yardstick against which team performance is measured.

- *Identify success criteria.*   The team must define what constitutes success and put it in writing. Remember, in a total quality setting, success should be defined in terms of the customer—internal and external. This means that team members must understand the needs and expectations of its customers before identifying success criteria.

- *Be action centered.*   For every success criterion, the team should develop an action statement or plan that specifies exactly what must be done to satisfy the criterion, the time frame within which it must be done, and by whom.

- *Establish the ground rules.*   The team needs to decide how it will operate. Team leaders should work to achieve consensus on such issues as these:

    Calling meetings only when necessary
    Making sure all team members come to meetings well briefed and fully prepared
    Determining how much time to allocate for agenda items
    Encouraging participants to be brief
    Determining who will serve as the recorder during meetings
    Deciding how and when to subdivide into subgroups
    Keeping disturbances and interruptions out of meetings
    Finishing an agenda item before moving on to the next item
    Allowing time for informed interaction among members before and/or after meetings

- *Share information.*   Information should be shared freely inside and outside the team. Communication is a fundamental element of total quality; everyone should know what is going on. Teams do not operate in a vacuum. They are all part of a larger team, the organization. Keeping everyone informed of team activities will eliminate idle, nonproductive, and typically inaccurate speculation.

- *Cultivate team unity.*   During the Summer Olympic Games one year the United States fielded what many sports experts and fans thought was the best basketball team ever assembled. Aptly named, the Dream Team won a gold medal with ease. But in spite of the enormous talent of individual team members, the gold medal was really won because a group of incredibly talented athletes developed a sense of unity, a team identity. As a result, they put aside individual goals, left their egos in the locker room, and played as a team—each supporting the other and contributing to the team's score rather than his own individual statistics. This does not mean they gave up their individual identities completely. On the contrary, the more team unity grew, the more supportive the team became of individual performances.

Team Excellence and Performance

Teamwork is not a magic cure-all. Poorly run teams can do more damage to an organization's performance and corresponding competitiveness than having no teams at all. For this reason, it is critical that excellence in team performance be an overriding goal of the organization.

Dennis King recommends the following strategies, which he calls the Ten Team Commandments:[6]

- *Interdependence.*   Team members should be mutually dependent on each other for information, resources, task accomplishment, and support. Interdependence is the glue that will hold a team together.
- *Stretching tasks.*   Teams need to be challenged. Responding to a challenge as a team builds esprit de corps and instills pride and unity.
- *Alignment.*   An aligned team is one in which all members not only share a common mission but are willing to put aside individualism to accomplish it.
- *Common language.*   Teams often consist of members from different departments (manufacturing, marketing, accounting, etc.). These different departments typically have their own indigenous languages that may be foreign to people from other departments. Consequently, it is important for team leaders to ensure that department-specific terms and phrases are used minimally and fully explained in common terms when used.
- *Trust/respect.*   For team members to work well together, there must be trust and respect. Time and effort spent building trust and respect among team members is time invested well.
- *Shared leadership/followership.*   Some group members tend to emerge as more vocal, while others sit back and observe. If group leaders are allowed to dominate, the group will not achieve its full potential. A better approach is to draw out the special talents of each individual group member so that leadership and followership are shared.
- *Problem-solving skills.*   Time invested in helping group members become better problem solvers is time well spent. Much of the business of groups is problem solving.
- *Confrontation-/conflict-handling skills.*   Human conflict is inevitable in a high-pressure, competitive workplace. Even the best teams and closest families have disagreements. Learning to disagree without being disagreeable and to air disagreement openly and frankly—attacking ideas, issues, and proposed solutions without attacking the people proposing them—are critical skills in a total quality environment.
- *Assessment/action.*   Assessment is a matter of asking and answering the question "How are we doing?" The yardsticks for answering this question are the group's mission statement and corresponding action plan. The action plan contains goals, objectives, timetables, and assignments of responsibility. By monitoring these continually, group members can assess how the group is doing.
- *Celebration.*   An effective team reinforces its successes by celebrating them. Recognition of a job well done can motivate team members to work even harder and smarter to achieve its next goal.

BUILDING TEAMS AND MAKING THEM WORK

Part of building a successful team is choosing team members wisely. This section describes strategies for selecting team members, naming officers (or otherwise assigning responsibility), creating a mission statement, and developing collegial relations among team members.

Makeup and Size of Teams

Teams should be composed of those people who are most likely to be able to satisfy the team's mission efficiently and effectively. The appropriate makeup of a team depends in part on the type of team in question (whether it is departmental, process improvement, or task force/project oriented). Departmental teams such as quality circles are made up of the employees of a given department. However, process improvement teams and task forces typically cross departmental lines.

The membership of such teams should be open to any level of employee—management, supervision, and hourly wage earners. A good rule of thumb is that the greater the mix the better. According to Perry L. Johnson:

> The bigger the mix, the better the results. As for the size of the team, we want a group large enough to guarantee a mix of people and opinions, yet small enough to make meetings comfortable, productive, and brief. So, what's the right number? No fewer than six, no more than twelve. Eight or nine is just about right.[7]

Choosing Team Members

When putting together a team, the first step is to identify all potential team members. This is important because there will often be more potential team members than the number of members actually needed (maximum of 12 members). After the list has been compiled, volunteers can be solicited and actual team members selected from among those who volunteer. However, care should be taken to ensure a broad mix, as set forth in the previous section. This rule should be adhered to even if there are no volunteers and team members must be drafted. The more likely case is that there will be more volunteers than openings on most teams. Johnson recommends handling this by periodically rotating the membership, preferably on an alternating basis so that the team always includes both experienced and new members.[8]

Responsibilities of Team Leaders

Most teams will have members who are managers, supervisors, and hourly employees. However, it should not be assumed that the highest-level manager will automatically be the team leader. Correspondingly, it should not be assumed that the most junior hourly worker cannot be the team leader. The team should select its own leader.

The first step in selecting a team leader is to develop an understanding of the role and responsibilities of this person. Beyond the description set forth, Scholtes lists the following as responsibilities of team leaders:[9]

- Serve as the official contact between the team and the rest of the organization.
- Serve as the official record keeper for the team. Records include minutes, correspondence, agendas, and reports. Typically, the team leader will appoint a recorder to take minutes during meetings. However, the team leader is still responsible for distributing and filing minutes.
- Serve as a full-fledged team member, but with care to avoid dominating team discussions.
- Implement team recommendations that fall within the team leader's realm of authority, and work with higher management to implement those that fall outside it.

Creating the Team's Mission Statement

After a team has been formed, a team leader selected, a reporter appointed, and a quality advisor assigned, the team is ready to draft its mission statement. This is a critical step in the life of a team. The mission statement explains the team's reason for being. A mission statement is written in terms that are broad enough to encompass all it will be expected to do but specific enough that progress can be easily measured. The following sample mission statement meets both of these criteria:

> The purpose of this team is to reduce the time between when an order is taken and when it is filled while simultaneously improving the quality of products shipped.

This statement is broad enough to encompass a wide range of activities and to give team members room within which to operate. The statement does not specify how much throughput time will be reduced nor how much quality will be improved. The level of specificity comes in the goals set by the team (e.g., reduce throughput time by 15% within 6 months; improve the customer satisfaction rate to 100% within 6 months). Goals follow the mission statement and explain it more fully in quantifiable terms.

This sample mission statement is written in broad terms, but it is specific enough that team members know they are expected to simultaneously improve both productivity and quality. It also meets one other important criterion: simplicity. Any employee could understand this mission statement. It is brief, to the point, and devoid of all esoteric nonessential verbiage.

Team leaders should keep these criteria in mind when developing mission statements: broadness, appropriate specificity, and simplicity. A good mission statement is a tool for communicating the team's purpose—within the team and throughout the organization—not a device for confusing people or an opportunity to show off literary dexterity.

Developing Collegial Relationships

A team works most effectively when individual team members form positive, mutually supportive peer relationships. These are collegial relationships, and they can be the difference between a high-performance team and a mediocre team. Scholtes recommends the following strategies for building collegial relationships among team members:[10]

■ Help team members understand the importance of honesty, reliability, and trustworthiness. Team members must trust each other and know that they can count on each other.

■ Help team members develop mutual confidence in their work ability.

■ Help team members understand the pressures to which other team members are subjected. It is important for team members to be supportive of peers as they deal with the pressures of the job.

These are the basics. Competence, trust, communication, and mutual support are the foundation on which effective teamwork is built. Any resources devoted to improving these factors is an investment well made.

Promoting Diversity in Teams

The American workplace has undergone an unprecedented transformation. Formerly dominated by young to middle-aged white males, the workplace now draws from a labor pool dominated by women and minorities. This means that today's employees come from a variety of different cultures and backgrounds. Consequently, they are likely to have different values and different outlooks. This situation can be good or bad, depending on how it is handled.

Handling diversity in a way that makes it a strength has come to be known as managing diversity. According to Sharon Nelton:

> Managing diversity meant, and still means, fostering an environment in which workers of all kinds—men, women, white, disabled, homosexual, straight, elderly—can flourish and, given opportunities to reach their full potential and contribute at the highest level, can give top performance to a company.[11]

When diversity is properly managed, the glass ceiling that exists in some workplaces can be eliminated. The glass ceiling consists of "artificial barriers that prevent qualified minorities and women from advancing into middle and senior levels of management."[12] By working together in well-managed teams that include women and men, young and old, minorities and nonminorities, employees can learn how to realize the full potential of diversity. Diversity in teamwork can be promoted by applying the following strategies:

■ *Continually assess circumstances.*   Is communication among diverse team members positive? Do bias and stereotyping exist among team members? Do minorities and nonminorities with comparable jobs and qualifications earn comparable wages? Factors that might undermine harmonious teamwork should be anticipated, identified, and dealt with.

■ *Give team members opportunities to learn.*   Humans naturally tend to distrust people who are different, whether the differences are attributed to gender, culture, age, race, or any other factor. Just working with people who are different can help overcome this unfortunate but natural human tendency. However, it usually takes more than just working together to break down barriers and turn a diverse group of employees into a mutually supportive, complementary team in which the effectiveness of the whole is greater than the sum of its parts. Education and training aimed at promoting sensitivity to and appreciation of human differences should be provided. Such training should also help team members overcome the stereotypical assumptions that society in general seems to promote.

For metal to have optimum strength and resiliency characteristics, it must be alloyed with other metals. High-performance, space-age metals are all mixtures of several different component metals, each different from the others and each possessing its own desirable characteristics. In the modern workplace, this analogy can be applied to the team. Diverse employees, properly managed and trained, can make high-performance, world-class teams.

FOUR-STEP APPROACH TO TEAM BUILDING

Effective team building is a four-step process:

1. Assess
2. Plan
3. Execute
4. Evaluate

To be a little more specific, the team-building process proceeds along the following lines: (a) assess the team's developmental needs (e.g., its strengths and weaknesses), (b) plan team-building activities based on the needs identified, (c) execute the planned team-building activities, and (d) evaluate results. The steps are spelled out further in the next sections.

Assessing Team Needs

If you were the coach of a baseball team about which you knew very little, what is the first thing you would want to do? Most coaches in such situations want to begin by assessing the abilities of their new team. Can we hit? Can we pitch? Can we field? What are our weaknesses? What are our strengths? With these questions answered, the coach will know how best to proceed with team-building activities.

This same approach can be used in the workplace. A mistake commonly made by organizations is beginning team-building activities without first assessing the team's developmental needs. Resources are often limited in organizations. Consequently, it is important to use them as efficiently and effectively as possible. Organizations that begin team-building activities without first assessing strengths and weaknesses run the risk of wasting resources in an attempt to strengthen characteristics that are already strong, while at the same time overlooking characteristics that are weak.

For workplace teams to be successful, they should have at least the following characteristics:

■ Clear direction that is understood by all members

■ "Team players" on the team

■ Fully understood and accepted accountability measures

Figure 7–1 is a tool that can be used for assessing the team-building needs of workplace teams. It consists of criteria arranged in three broad categories: *direction and understanding, characteristics of team members*, and *accountability*. Individual team members record their perceptions of the team's performance and abilities relative to the specific criteria in each category. The highest score possible for each criterion is 6; the lowest score possible, 0. The team score for each criterion is found by adding the scores of individual members for that criterion and dividing by the number of team members. For example, a four-person team might produce the following score on a given criterion:

Instructions

To the left of each item is a blank for recording your perception regarding that item. For each item, record your perception of how well it describes your team. Is the statement *Completely True* (CT), *Somewhat True* (ST), *Somewhat False* (SF), or *Completely False* (CF)? Use the following numbers to record your perception.

CT = 6
ST = 4
SF = 2
CF = 0

Direction and Understanding

_____ 1. The team has a clearly stated mission.

_____ 2. All team members understand the mission.

_____ 3. All team members understand the scope and boundaries of the team's charter.

_____ 4. The team has a set of broad goals that support its mission.

_____ 5. All team members understand the team's goals.

_____ 6. The team has identified specific activities that must be completed in order to accomplish team goals.

_____ 7. All team members understand the specific activities that must be completed in order to accomplish team goals.

_____ 8. All team members understand projected timeframes, schedules, and deadlines relating to specific activities.

Characteristics of Team Members

_____ 9. All team members are open and honest with each other all the time.

_____ 10. All team members trust each other.

_____ 11. All team members put the team's mission and goals ahead of their own personal agendas all of the time.

_____ 12. All team members are comfortable that they can depend on each other.

_____ 13. All team members are enthusiastic about accomplishing the team's mission and goals.

_____ 14. All team members are willing to take responsibility for the team's performance.

_____ 15. All team members are willing to cooperate in order to get the team's mission accomplished.

Figure 7–1
Team-Building Needs Assessment

_____ 16. All team members will take the initiative in moving the team toward its final destination.

_____ 17. All team members are patient with each other.

_____ 18. All team members are resourceful in finding ways to accomplish the team's mission in spite of difficulties.

_____ 19. All team members are punctual when it comes to team meetings, other team activities, and meeting deadlines.

_____ 20. All team members are tolerant and sensitive to the individual differences of team members.

_____ 21. All team members are willing to persevere when team activities become difficult.

_____ 22. The team has a mutually supportive climate.

_____ 23. All team members are comfortable expressing opinions, pointing out problems, and offering constructive criticism.

_____ 24. All team members support team decisions once they are made.

_____ 25. All team members understand how the team fits into the overall organization/big picture.

Accountability

_____ 26. All team members know how team progress/performance will be measured.

_____ 27. All team members understand how team success is defined.

_____ 28. All team members understand how ineffective team members will be dealt with.

_____ 29. All team members understand how team decisions are made.

_____ 30. All team members know their respective responsibilities.

_____ 31. All team members know the responsibilities of all other team members.

_____ 32. All team members understand their authority within the team and that of all other team members.

_____ 33. All team goals have been prioritized.

_____ 34. All specific activities relating to team goals have been assigned appropriately and given projected completion dates.

_____ 35. All team members know what to do when unforeseen inhibitors impede progress.

Figure 7–1 continued

Team member 1:   4
Team member 2:   2
Team member 3:   2
Team member 4:   4
                  $\overline{12}$: $12 \div 4 = 3$ (team average score)

The lower the team score for each criterion, the more work needed on that criterion.

Team-building activities should be developed and executed based on what is revealed by this assessment. Activities should be undertaken in reverse order of the assessment scores (e.g., lower scores first, higher scores last). For example, if the team score for criterion I (clearly stated mission) is the lowest score for all the criteria, the first team-building activity would be the development of a mission statement.

Planning Team-Building Activities

Team-building activities should be planned around the results of the needs assessment conducted in the previous step. Consider the example of a newly chartered team. The highest score for a given criteria in Figure 7–1 is 6. Consequently, any team average score less than 6 indicates a need for team building relating to the criterion in question. The lower the score, the greater the need.

For example, say the team in question had an average score of 3 for criterion 2 ("All team members understand the mission"). Clearly, part of the process of building this team must be explaining the team's mission more clearly. A team average score of 3 on this issue indicates that some members understand the mission and some don't. This solution might be as simple as the responsible manager or team leader sitting down with the team, describing the mission, and responding to questions from team members.

On the other hand, if the assessment produces a low score for criterion 9 ("All team members are open and honest with each other all the time"), more extensive trust-building activities may be needed. In any case, what is important in this step is to (a) plan team-building activities based on what is learned from the needs assessment and (b) provide team-building activities in the priority indicated by the needs assessment, beginning with the lowest scores.

Executing Team-Building Activities

Team-building activities should be implemented on a just-in-time basis. A mistake made by many organizations that are interested in implementing total quality is rushing into team building. All employees are given teamwork training, even those who are not yet part of a chartered team. Like any kind of training, teamwork training will be forgotten unless it is put to immediate use. Consequently, the best time to provide teamwork training is after a team has been formed and given its charter. In this way, team members will have opportunities to apply what they are learning immediately.

Team building is an ongoing process. The idea is to make a team better and better as time goes by. Consequently, basic teamwork training is provided as soon as a team is chartered. All subsequent team-building activities are based on the results of the needs assessment and planning process.

Evaluating Team-Building Activities

If team-building activities have been effective, weaknesses pointed out by the needs assessment process should have been strengthened. A simple way to evaluate the effectiveness of team-building activities is to readminister the appropriate portion of the needs assessment document. The best approach is to reconstitute the document so that it contains the relevant criteria only. This will focus the attention of team members on the specific targeted areas.

If the evaluation shows that sufficient progress has been made, nothing more is required. If not, additional team-building activities are needed. If a given team-building activity appears to have been ineffective, get the team together and discuss it. Use the feedback from team members to identify weaknesses and problems and use the information to ensure that team-building activities become more effective.

TEAMS ARE NOT BOSSED—THEY ARE COACHED

If employees are going to be expected to work together as a team, managers and supervisors have to realize that teams are not bossed—they are coached. Team leaders, regardless of their respective titles (manager, supervisor, etc.) need to understand the difference between bossing and coaching. Bossing, in the traditional sense, involves planning work, giving orders, monitoring programs, and evaluating performance. Bosses approach the job from an "I'm in charge—do as you are told" perspective.

Coaches, on the other hand, are facilitators of team development and continually improved performance. They approach the job from the perspective of leading the team in such a way that it achieves peak performance levels on a consistent basis. This philosophy is translated into everyday behavior as follows:

■ Coaches give their teams a clearly defined charter.
■ Coaches make team development and team building a constant activity.
■ Coaches are mentors.
■ Coaches promote mutual respect between themselves and team members and among team members.
■ Coaches make human diversity within a team a plus.

Clearly Defined Charter

One can imagine a basketball, soccer, or track coach calling her team together and saying, "This year we have one overriding purpose—to win the championship." In one simple statement this coach has clearly and succinctly defined the team's charter. All team members now know that everything they do this season should be directed at winning the championship. The coach didn't say the team would improve its record by 25 points, improve its standing in the league by two places, or make the playoffs, all of which would be worthy missions. This coach has a higher vision. This year the team is going for the championship. Coaches of work teams should be just as specific in explaining the team's mission to team members.

Team Development/Team Building

The most constant presence in an athlete's life is practice. Regardless of the sport, athletic teams practice constantly. During practice, coaches work on developing the skills of individual team members and the team as a whole. Team development and team-building activities are ongoing forever. Coaches of work teams should follow the lead of their athletic counterparts. Developing the skills of individual team members and building the team as a whole should be a normal part of the job—a part that takes place regularly, forever.

Mentoring

Good coaches are mentors. This means they establish a helping, caring, nurturing relationship with team members. Developing the capabilities of team members, improving the contribution individuals make to the team, and helping team members advance their careers are all mentoring activities. According to Gordon F. Shea, effective mentors help team members in the following ways:[13]

■ Developing their job-related competence
■ Building character
■ Teaching them the corporate culture
■ Teaching them how to get things done in the organization
■ Helping them understand other people and their viewpoints
■ Teaching them how to behave in unfamiliar settings or circumstances
■ Giving them insight into differences among people
■ Helping them develop success-oriented values

Mutual Respect

It is important for team members to respect their coach, for the coach to respect his or her team members, and for team members to respect each other. According to Robert H. Rosen, "Respect is composed of a number of elements that, like a chemical mixture, interact and bond together."[14]

- *Trust made tangible.*   Trust is built by (a) setting the example, (b) sharing information, (c) explaining personal motives, (d) avoiding both personal criticisms and personal favors, (e) handing out sincere rewards and recognition, and (f) being consistent in disciplining.

- *Appreciation of people as assets.*   Appreciation for people is shown by (a) respecting their thoughts, feelings, values, and fears; (b) respecting their desire to lead and follow; (c) respecting their individual strengths and differences; (d) respecting their desire to be involved and to participate; (e) respecting their need to be winners; (f) respecting their need to learn, grow, and develop; (g) respecting their need for a safe and healthy workplace that is conducive to peak performance; and (h) respecting their personal and family lives.

- *Communication that is clear and candid.*   Communication can be made clear and candid if coaches will do the following: (a) open their eyes and ears—observe and listen; (b) say what they want, say what they mean (be tactfully candid); (c) give feedback constantly and encourage team members to follow suit; and (d) face conflict within the team head-on—that is, don't let resentment among team members simmer until it boils over—handle it now.

- *Ethics that are unequivocal.*   Ethics can be made unequivocal by (a) working with the team to develop a code of ethics; (b) identifying ethical conflicts or potential conflicts as early as possible; (c) rewarding ethical behavior; (d) disciplining unethical behavior, and doing so consistently; and (e) before bringing in new team members, make them aware of the team's code of ethics. In addition to these strategies, the coach should set a consistent example of unequivocal ethical behavior.

- *Team members are assets.*   Professional athletes in the United States are provided the best medical, health, and fitness services in the world. In addition, they practice and perform in an environment that is as safe and healthy as it can be made. Their coaches insist on these conditions because they understand that the athletes are invaluable resources. Their performance determines the ultimate success or failure of the organization. Coaches of work teams should take a similar approach. To protect their assets (team members), coaches can apply the following strategies: (a) form a partnership between the larger organization and the team to promote healthy habits and control; (b) encourage monitoring and screening of high-risk conditions such as high blood pressure, high cholesterol, and cancer; (c) promote nonsmoking; (d) encourage good nutrition and regular exercise; (e) organize classes, seminars, or workshops on such subjects as HIV/AIDS, prenatal care, stress management, stroke and heart attack prevention, workplace safety, nutrition, and ergonomics; (f) encourage higher management to establish an employee-assistance plan (EAP); and (g) stress important topics such as accident prevention and safe work methods.

Human Diversity

Human diversity is a plus. Sports and the military have typically led American society in the drive for diversity, and both have benefited immensely as a result. To list the contributions to either sports or the military made by people of different genders, races, religions, and so on, would be a task of gargantuan portions. Fortunately, leading organizations in the United States have followed the positive example set by sports and the military. The smart ones have learned that most of the growth in the workplace will be among women, minorities, and immigrants. These people will bring new ideas and different perspectives, precisely what an organization needs to stay on the razor's edge of competitiveness. However, in spite of steps already taken toward making the American workplace both diverse and harmonious, wise coaches understand that people—consciously and unconsciously—tend to erect barriers between themselves and people who are different from them. This tendency can quickly undermine that trust and co-hesiveness upon which teamwork is built. To keep this from happening, coaches can do the following:

■ *Conduct a cultural audit.*   Identify the demographics, personal characteristics, cultural values, and individual differences among team members.

■ *Identify the specific needs of different groups.*   Ask women, ethnic minorities, and older workers to describe the unique inhibitors they face. Make sure all team members understand these barriers, then work together as a team to eliminate, overcome, or accommodate them.

■ *Confront cultural clashes.*   Wise coaches meet conflict among team members head-on and immediately. This approach is particularly important when the conflict is based on issues of diversity. Conflicts that grow out of religious, cultural, ethnic, age-, and/or gender-related issues are more potentially volatile than everyday disagreements over work-related issues. Consequently, conflict that is based on or aggravated by human differences should be confronted promptly. Few things will polarize a team faster than diversity-related disagreements that are allowed to fester and grow.

■ *Eliminate institutionalized bias.*   A company whose workforce had historically been predominantly male now has a workforce in which women are the majority. However, the physical facility still has 10 men's restrooms and only two for women. This imbalance is an example of institutionalized bias. Teams may find themselves unintentionally slighting members, simply out of habit or tradition. This is the concept of *discrimination by inertia*. It happens when the demographics of a team changes but its habits, traditions, procedures, and work environment do not.

An effective way to eliminate institutional bias is to circulate a blank notebook and ask team members to record—without attribution—instances and examples of institutional bias. After the initial circulation, repeat the process periodically. The coach can use the input collected to help eliminate institutionalized bias. By collecting input directly from team members and acting on it promptly, coaches can ensure that discrimination by inertia is not creating or perpetuating quiet but debilitating resentment.

HANDLING CONFLICT IN TEAMS

The following conversation took place in a meeting the authors once attended: A CEO had called together employees in his company to deal with issues that were disrupting work. Where the company wanted teamwork, it was getting conflict. Where it wanted mutual cooperation, it was getting bickering. The conversation started something like this:

CEO:	We all work for the same company, don't we?
Employees:	[Nods of agreement.]
CEO:	We all understand that we cannot do well unless the company does well don't we?
Employees:	[Nods of agreement.]
CEO:	Then we want the company to do well, don't we?
Employees:	[Nods of agreement.]
CEO:	Then we are all going to work together toward the same goal, aren't we?
Employees:	[Silence. All employees stared uncomfortably at the floor.]

This CEO made a common mistake. He thought that employees would automatically work together as a team because this approach is so obviously the right thing to do. In other words, just give employees a chance and explain things to them and they'll work together. James R. Lucas calls this belief a "fatal illusion."[15] He lists the following reasons that people might not work well in teams—reasons that, in turn, can lead to conflict:[16]

■ Because we live in an age of rapid change, employees can conclude that the only person they can trust is themselves.

■ Employees who are too proud or too broken and wounded by others might resist relying on others to help achieve their goals.

■ Employees steeped in the traditions of *rugged individuality* and *competition is king* can feel that cooperation is unfitting for a vigorous person or organization.

In addition to these personal inhibitors of teamwork and promoters of conflict, there is the *example* issue. Organizations that espouse teamwork among employees but clearly are not good team players themselves are setting an example that works against teamwork. Poor teamwork on the part of an organization will manifest itself in either or both of the following ways: (a) treating suppliers poorly while advocating a partnership and (b) treating customers poorly while advocating customer satisfaction.

If organizations want employees to be team players, they must set a positive example of teamwork. If organizations want employees to resolve team conflicts in a positive manner, they must set an example of resolving supplier and customer conflicts in a positive manner.

Resolution Strategies for Team Conflicts

Conflict will occur in even the best teams. Even when all team members agree on a goal, they can still disagree on how best to accomplish it. Lucas recommends the following strategies for preventing and resolving team conflict:[17]

■ Plan and work to establish a culture where individuality and dissent are in balance with teamwork and cooperation.

■ Establish clear criteria for deciding when decisions will be made by individuals and when they will be made by teams.

■ Don't allow individuals to build personal empires or to use the organization to advance personal agendas.

■ Encourage and recognize individual risk-taking behavior that breaks the organization out of unhelpful habits and negative mental frameworks.

■ Encourage healthy, productive competition, and discourage unhealthy, counterproductive competition.

■ Recognize how difficult it can be to ensure effective cooperation, and spend the energy necessary to get just the right amount of it.

■ Value constructive dissent, and encourage it.

■ Assign people of widely differing perspectives to every team or problem.

■ Reward and recognize both dissent and teamwork when they solve problems.

■ Reevaluate the project, problem, or idea when no dissent or doubt is expressed.

■ Avoid hiring people who think they don't need help, who don't value cooperation, or who are driven by the desire to be accepted.

■ Ingrain into new employees the need for balance between the concept of cooperation and constructive dissent.

■ Provide ways for employees to say what no one wants to hear.

■ Realistically and regularly assess the ability and willingness of employees to cooperate effectively.

- Understand that some employees are going to clash, so determine where this is happening and remix rather than waste precious organizational energy trying to get people to like each other.
- Ensure that the organization's value system and reward/recognition systems are geared toward cooperation with constructive dissent rather than dog-eat-dog competition or cooperating at all costs.
- Teach employees how to manage both dissent (not let it get out of hand) and agreement (not let them get out of hand).
- Quickly assess whether conflict is healthy or destructive, and take immediate steps to encourage the former and resolve or eliminate the latter.

REWARDING TEAM AND INDIVIDUAL PERFORMANCE

An organization's attempts to institutionalize teamwork will fail unless it includes implementation of an appropriate compensation system; said another way, if you want teamwork to work, make it pay. This does not mean that employees are no longer compensated as individuals. Rather, the most successful compensation systems combine both individual and team pay.

This matter is important because few employees work exclusively in teams. A typical employee, even in the most team-oriented organization, spends a percentage of his or her time involved in team participation and a percentage involved in individual activities. Even those who work full-time in teams have their individual responsibilities that are carried out on behalf of the team.

Consequently, the most successful compensation systems have the following components: (a) base compensation for the individual; (b) individual incentive compensation; and (c) team incentive compensation. With such a system, all employees receive their traditional individual base pay. Then there are incentives that allow employees to increase their income by surpassing goals set for their individual performance. Finally, other incentives are based on team performance. In some cases the amount of team compensation awarded to individual team members is based on their individual performance within the team or, in other words, on the contribution they made to the teams' performance.

An example of this approach can be found in the world of professional sports. All baseball players in both the National and American Leagues receive a base amount of individual compensation. Most also have a number of incentive clauses in their contracts to promote better individual performance. Team-based incentives are offered if the team wins the World Series or the league championship. When this happens, the players on the team divide the incentive dollars into shares. Every member of the team receives a certain number of shares based on his perceived contribution to the team's success that year.

Figure 7–2 is a model that can be used for establishing a compensation system that reinforces both team and individual performance. The first step in this model involves deciding what performance outcomes will be measured (individual and team outcomes). Step 2 involves how the outcomes will be measured. What types of data will tell the story? How can these data be collected? How frequently will the performance measurements be made? Step 3 involves deciding what types of rewards will be offered (monetary, nonmonetary, or a combination of the two). This is the step in which rewards are organized into levels that correspond to levels of performance so that the reward is in proportion to the performance.

The issue of proportionality is important when designing incentives. If just barely exceeding a performance goal results in the same reward given for substantially exceeding it, just barely is what the organization will get. If exceeding a goal by 10% results in a 10% bonus, then exceeding it by 20% should result in a 20% bonus, and so

Figure 7–2
Model for Developing a Team
and Individual Compensation
System

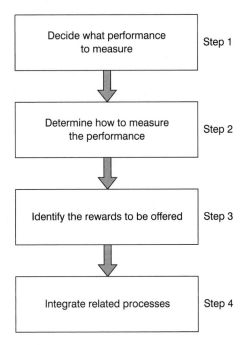

on. Proportionality and fairness are characteristics that employees scrutinize with care when examining an incentive formula. Any formula that is seen to be unfair or disproportionate will not have the desired result.

The final step in the model in Figure 7–2 involves integrating the compensation system with other performance-related processes. These systems include performance appraisal, the promotion process, and staffing. If teamwork is important, one or more criteria relating to teamwork should be included in the organization's performance appraisal process.

Correspondingly, the employee's ratings on the teamwork criteria in a performance appraisal should be considered when making promotion decisions. An ineffective team player should not be promoted in an organization that values teamwork. Other employees will know, and teamwork will be undermined. Finally, during the selection process, applicants should be questioned concerning their views on teamwork. It makes no sense for an organization that values teamwork to hire new employees who, during their interview, show no interest in or aptitude for teamwork.

Nonmonetary Rewards

A common mistake made when organizations first attempt to develop incentives is thinking that employees will respond only to dollars in a paycheck. In fact, according to Douglas G. Shaw and Craig Eric Schneier, nonmonetary rewards can be even more effective than actual dollars.[18] Widely used nonmonetary rewards that have proven to be effective include the following: movie tickets, gift certificates, time off, event tickets, free attendance at seminars, getaway weekends for two, airline tickets, and prizes such as electronic and/or household products.

Different people respond to different incentives. Consequently, what will work can be difficult to predict. A good rule of thumb to apply when selecting nonmonetary incentives is "Don't assume—ask." Employees know what appeals to them. Before investing in nonmonetary incentives, organizations should survey their employees. List as many different potential nonmonetary rewards as possible and let employees rate them. In addition, set up the incentive system so that employees, to the extent possible, are able to select the reward that appeals to them. For example, employees who exceed performance goals (team or individual) by 10% should be allowed to select from among several equally valuable rewards on the "10% Menu." Where one employee might enjoy dinner tickets for two, another might be more motivated by tickets to a sporting event. The better an incentive program is able to respond to individual preferences, the better it will work.

RECOGNIZING TEAMWORK AND TEAM PLAYERS

One of the strongest human motivators is *recognition*. People don't just want to be recognized for their contributions; they *need* to be recognized. The military applies this fact very effectively. The entire system of military commendations and decorations (medals) is based on the positive human response to recognition. No amount of pay could compel a young soldier to perform the acts of bravery that are commonplace in the history of the United States military. But the recognition of a grateful nation continues to spur men and women on to incredible acts of valor every time our country is involved in an armed conflict. There is a lesson here for nonmilitary organizations.

The list of methods for recognizing employees goes on ad infinitum. There is no end to the ways that the intangible concept of employee appreciation can be demonstrated. For example, writing in the *Quality Digest*, Bob Nelson recommends the following methods:[19]

■ Write a letter to the employee's family telling them about the excellent job the employee is doing.

■ Arrange for a senior-level manager to have lunch with the employee.

■ Have the CEO of the organization call the employee personally (or stop by in person) to say, "Thanks for a job well done."

■ Find out what the employee's hobby is and publicly award him or her a gift relating to that hobby.

■ Designate the best parking space in the lot for the "Employee of the Month."

■ Create a "Wall of Fame" to honor outstanding performance.

These examples are provided to trigger ideas but are only a portion of the many ways that employees can be recognized. Every individual organization should develop its own locally tailored recognition options. When doing so, the following rules of thumb will be helpful:

■ Involve employees in identifying the types of recognition activities to be used. Employees are the best judge of what will motivate them.

■ Change the list of recognition activities periodically. The same activities used over and over for too long will go stale.

■ Have a variety of recognition options for each level of performance. This will allow employees to select the type of reward that appeals to them the most.

ENDNOTES

1. Perry L. Johnson, Rob Kantner, and Marcia A. Kikora, *TQM Team-Building and Problem-Solving* (Southfield, MI: Perry Johnson, Inc., 1990), 1–1.
2. Johnson, Kantner, and Kikora, 1–2.
3. Peter R. Scholtes, *The Team Handbook* (Madison, WI: Joiner Associates, 1992), 6–1.
4. Scholtes, 6–1, 6–2.

5. Mary Walsh Massop, "Total Teamwork: How to Be a Member," in *Management for the 90's: A Special Report from SUPERVISORY MANAGEMENT* (Saranc Lake, NY: American Management Association, 1991), 8.

6. Dennis King, "Team Excellence," in *Management for the 90's: A Special Report from SUPERVISORY MANAGEMENT* (Saranac Lake, NY: American Management Association, 1991), 16–17.

7. Johnson, Kantner, and Kikora, 2–1.

8. Johnson, Kantner, and Kikora, 2–2, 2–3.

9. Scholtes, 3–9, 3–10.

10. Scholtes, 3–8.

11. Sharon Nelton, "Winning with Diversity," *Nation's Business* (September 1992): 19.

12. Nelton, 19.

13. Gordon F. Shea, *Mentoring* (New York: American Management Association, 1994), 49–50.

14. Robert H. Rosen, *The Healthy Company* (New York: Perigee Books, Putnam, 1991), 24.

15. James R. Lucas, *Fatal Illusions* (New York: AMACOM, American Management Association, 1996), 155.

16. Lucas, 159–160.

17. Lucas, 160–161.

18. Douglas G. Shaw and Craig Eric Schneier, "Team Measurements and Rewards: How Some Companies Are Getting It Right," *Human Resource Planning* 18(3) (1995): 39.

19. Bob Nelson, "Secrets of Successful Employee Recognition," *Quality Digest* 16(8) (August 1996): 29.

Training

One of the most fundamental elements of total quality is the ongoing development of personnel, which means training and learning. This chapter provides readers with the information needed to justify, provide, and evaluate training and learning.

OVERVIEW OF TRAINING AND LEARNING

It is common to hear the terms *education, training*, and *learning* used interchangeably in discussions of employee development. Although common practice is to use the term *training* for the sake of convenience, modern managers should be familiar with distinctions. For purposes of this book, training is defined as follows:

> Training is an organized, systematic series of activities designed to enhance an individual's work-related knowledge, skills, and understanding and/or motivation.

Training can be distinguished from *education* by its characteristics of practicality, specificity, and immediacy. Training should relate specifically to the job performed by those being trained, or it should have immediate practical application on the job. *Education* is a broader term; training is a subset of education. Also, education tends to be more philosophical and theoretical and less practical than training.

The purpose of both education and training is learning. In an educational setting the learning will tend to be more theoretical, whereas in a training setting it will be more practical. However, with both, understanding is implicit in learning. Whether the point is to have the learner understand *why* or *how to*, the point is still to have the learner understand. Understanding is what allows an employee to become an innovator, initiative taker, and creative problem solver in addition to being an efficient and effective performer of his or her job.

RATIONALE FOR TRAINING

The rationale for training can be found in the need to compete. To survive in the modern marketplace, organizations must be able to compete globally. Companies that at one time competed only with their neighbors up the street now find themselves competing against companies from Europe, Asia, Central and South America, and the Pacific Rim. These companies are like the local high school track star who decides to try out for the Olympic team. Suddenly the competition is much more difficult, and it will be increasingly so at each successive level right up to the Olympic Games. If the athlete is able to make it that far, he or she will face the best athletes in the world. This is the situation in which modern business and industrial firms find themselves every day. Like the Olympic team that must have world-class athletes in order to win medals, these companies must have world-class employees to win the competition for market share.

Several factors combine to magnify the need for training. The most important of these are as follows:

- Poor quality of the existing labor pool
- Global competition
- Rapid and continual change
- Technology transfer problems
- Changing demographics

Benefits of Training

In spite of the fact that billions of dollars are spent on training each year, many employers still do not understand the role or benefits of training in the modern workplace. John Hoerr, writing in *Business Week*, paints a grim picture of the attitudes of U.S. employers toward training.[1] Hoerr cites a study conducted by the National Center on Education and the Economy that compares the education and training of workers in the United States with those of workers in competing countries.

This study concludes that less than 10% of U.S. companies use a flexible approach requiring better-trained workers as a way to improve productivity. This approach is standard practice in Japan, Germany, Denmark, and Sweden. In addition, less than 30% of U.S. firms have special training programs for women, minorities, and immigrants, in spite of the fact that over 80% of all new workers come from these groups. It is critical to the competitiveness of U.S. industry that employers understand the need for training that results from such factors as intense international competition, rapid and continual change, technology transfer problems, and changing demographics.

Tom Peters claims, "Our investment in training is a national disgrace."[2] According to Peters, 69% of the companies in the United States with 50 or more employees provide training for middle managers, and 70% train executive-level personnel. However, only 30% train skilled personnel. Peters contrasts this with the Japanese and Germans, who out-spend U.S. firms markedly in providing training for skilled personnel.

Modern managers should understand the benefits of workforce training and be able to articulate these benefits to higher management. Figure 8–1 is a checklist that summarizes the more important of these. As can be seen from the checklist, the benefits of training build upon themselves. For example, reducing the turnover rate will also contribute to improving safety. Increased safety will, in turn, help minimize insurance costs.

Employers who are not familiar with the benefits of training but are beginning to take an interest in providing it often debate the applicability or job relatedness of training. The argument is often made that "We will provide only the training that relates directly to the job." According to total quality pioneer W. Edwards Deming, focusing too intently on direct applicability is a mistake. Any kind of learning can benefit employees and employers alike in ways that cannot be predicted.

Figure 8–1
Checklist of Training Benefits

_____ Fewer Production Errors
_____ Increased Productivity
_____ Improved Quality
_____ Decreased Turnover Rate
_____ Lower Staffing Costs
_____ Improved Safety and Health
_____ Fewer Accidents
_____ Minimized Insurance Costs
_____ Increased Flexibility of Employees
_____ Better Response to Change
_____ Improved Communication
_____ Better Teamwork
_____ More Harmonious Employee Relations

TRAINING NEEDS ASSESSMENT

How do managers know what training is needed in their organizations? The answer is that many don't know. When compared with their competitors from other countries, U.S. companies appear to spend a great deal of money on the wrong kinds of training. It is a matter of emphasis. By placing the emphasis on management, employers are spending the bulk of their training dollars on those who organize and oversee the work rather than those who actually do it. This is akin to training the coaches instead of the players.

This is not to say that managers don't need ongoing training. In a total quality setting, every employee needs continual training. However, the keys to maximizing the return on training dollars are to place the emphasis on training those who need it most and ensure that the training provided is designed to promote the goals of the organization (quality, productivity, competitiveness).

Satisfying the first criterion is simply a matter of reversing the emphasis so that it is bottom-up in nature rather than top-down. Satisfying the second involves assessing training needs. Begin by asking two broad questions:

■ What knowledge, skills, and attitudes do our employees need to have to be world-class?

■ What knowledge, skills, and attitudes do our employees currently have?

The difference between the answers to these questions identifies an organization's training needs.

Managers may become involved in assessing training needs at two levels: the organizational level and the individual level. Managers who work closely enough with their team members can see firsthand on a daily basis their capabilities and those of the team as a whole. Observation is one method managers can use for assessing training needs. Are there specific problems that persist? Does an individual have problems performing certain tasks? Does work consistently back up at a given point in the process? These are indicators of a possible need for training that can be observed.

A more structured way to assess training needs is to ask employees to state their needs in terms of their job knowledge and skills. Employees know the tasks they must perform every day. They also know which tasks they do well, which they do not do well, and which they cannot do at all. A brainstorming session focusing on training needs is another method managers can use. Brainstorming is particularly effective in organizations where employees are comfortable speaking out as part of the continual improvement process.

The most structured approach managers can use to assess training needs is the *job task analysis survey*. With this approach, a job is analyzed thoroughly and the knowledge, skills, and attitudes needed to perform it are recorded. Using this information, a survey instrument is developed and distributed among employees who do the job in question. They respond by indicating which skills they have and which they need.

Before preparing a survey instrument, step back and take in the big picture. A common mistake is to focus too intently on the finite tasks of a job to the exclusion of the broader, less tangible requirements. For example, while doing a comprehensive job of breaking down technical process tasks, managers often overlook such criteria as teamwork skills, sensitivity to customer feedback (particularly internal customers), problem solving, and interpersonal skills. For this reason, it is a good idea to involve the employees who will be surveyed in the development of the survey instrument.

Another way to identify training needs is to ask employee groups to convene their own quality circles relating specifically to training. Employee-managed groups such as these that convene without managers and supervisors are often more open to admitting that there are training needs. They will also be less reticent to identify the training needs they think supervisors and managers have. The organization's suggestion system should also be used to identify training needs.

Converting Training Needs to Training Objectives

Having identified training needs, the next step is to write training objectives. This responsibility will fall in whole or in part to the manager. Some organizations have training personnel who can assist; others do not. In either case, managers in a total quality setting should be proficient in writing training objectives. The key to writing good training objectives lies in learning to be specific and to state objectives in behavioral terms. For example, suppose a need for training in the area of mathematics has been identified. He or she might write the following training objective:

> Employees will learn mathematics.

This training objective, as stated, lacks specificity and is stated in nonbehavioral terms. Mathematics is a broad concept. What is the need? Arithmetic? Algebra? Geometry? Trigonometry? All of these? To gain specificity, this objective must be broken into several objectives and restated.

To be stated in behavioral terms, these more specific objectives must explain what the employee should be able to do after completing the training. Behavioral objectives contain action verbs. The sample training objectives in Figure 8–2 are stated in behavioral terms, they are specific, and they are measurable. The more clearly training objectives are written, the easier it is to plan training to meet them. The training objectives in Figure 8–2 are all stated specifically and in behavioral terms. Taking the time to write objectives in this way will make it easier to provide and evaluate training.

---

1. Upon completion of this lesson, employees will be able to solve right triangles.
2. Upon completion of this lesson, employees will be able to apply the Law of Sines to the solution of triangles.
3. Upon completion of this lesson, employees will be able to apply the Law of Cosines to the solution of triangles.
4. Upon completion of this lesson, employees will be able to add, subtract, multiply, and divide decimal fractions.
5. Upon completion of this lesson, employees will be able to solve equations containing one unknown variable.

Figure 8–2
Sample Training Objectives

PROVIDING TRAINING

Ernest Boyer, president of the Carnegie Foundation for the Advancement of Teaching, sets forth four dimensions of the corporate learning enterprise.[3] Boyer credits Nell P. Eurich, a trustee of the Carnegie Foundation, with identifying the following dimensions of corporate education and training:

- In-house education is becoming more frequently used to meet the needs of industry and business including needs typically referred to as general education.
- Educational and training facilities owned and operated by corporations are becoming common. Motorola, Xerox, RCA, Rockwell International, and many other companies established their own education and training facilities.
- Degree-granting institutions that are corporate owned and operated and that are accredited by the same associations that accredit both public and private colleges and universities are becoming common. Many community colleges offer degree-granting partnership programs with industrial firms.
- The Satellite University through which instruction is provided by satellite to local downlink sites is becoming common. The National Technological University (NTU) of Fort Collins, Colorado, transmits instruction up through the master's degree to corporate classrooms nationwide.

Several different methods for providing training are available to organizations. All fall into one of the following three broad categories: internal approaches, external approaches, and partnership approaches. Regardless of the approach used, it is important to maximize training resources. Carolyn Wilson recommends five strategies for maximizing training resources:[4]

1. *Build in quality from the start.*   Take the time to do it right from the outset.
2. *Design small.*   Do not try to develop courses that are all things to all people. Develop specific activities around specific objectives.
3. *Think creatively.*   Do not assume that the traditional classroom approach is automatically best. Videotapes, interactive video, or one-on-one peer training may be more effective in some cases.
4. *Shop around.*   Before purchasing training services, conduct a thorough analysis of specific job training objectives. Decide exactly what you want and make sure the company you plan to deal with can provide it.
5. *Preview and customize.*   Never buy a training product (videotape, self-paced manual, or something else) without previewing it. If you can save by customizing a generic product, do so.

Internal Approaches

Internal approaches are those used to provide training on-site in the organization's facilities. These approaches include one-on-one training, on-the-job computer-based training, formal group instruction, and media-based instruction. *One-on-one training* involves placing a less skilled, less experienced employee under the instruction of a more skilled, more experienced employee. This approach is often used when a new employee is hired. It is also an effective way to prepare a replacement for a high-value employee who plans to leave or retire.

Computer-based training (CBT) has proven to be an effective internal approach. Over the years, it has continually improved so that now CBT is a widely used training method. It offers the advantages of being self-paced, individualized, and able to provide immediate and continual feedback to learners. Its best application is in developing general knowledge rather than in company-specific job skills. On-Line courses over the Internet are now widely available.

Formal *group instruction* in which a number of people who share a common training need are trained together is a widely used method. This approach might involve lectures, demonstrations, multimedia use, hands-on learning, question/answer sessions, role playing, and simulation.

Media-based instruction has become a widely used internal approach. Private training companies and major publishing houses produce an almost endless list of turnkey media-based training programs. The simplest of these might consist of a set of audiotapes. A more comprehensive package might include videotapes and workbooks. Interactive laser disk training packages that combine computer, video, and laser disk technology are also effective with the internal approach.

External Approaches

External approaches are those that involve enrolling employees in programs or activities provided by public institutions, private institutions, professional organizations, and private training companies. The two most widely used approaches are (a) enrolling employees in short-term training (a few hours to a few weeks) during work hours and (b) enrolling employees in long-term training such as a college course and paying all or part of the costs (i.e., tuition, books, fees). External approaches encompass training methods ranging from seminars to college courses. The external approach is typically used for developing broad, generic skills. However, some institutions will work with employers to develop customized courses.

Partnership Approaches

In recent years, community colleges, universities, and technical schools have begun to actively pursue partnerships with employers through which they provide customized training. These training partnerships combine some of the characteristics of the previous two approaches.

Customized on-site training provided cooperatively by colleges and private companies or associations have become very common. According to Nell Eurich of the Carnegie Foundation, the more than 1,200 community colleges in the United States have built extensive networks and alliances with business and industry.[5] Eurich cites the example of General Motors Corporation, which has contracts with 45 community colleges throughout the country to train service technicians. General Motors contributes to the partnership by training the instructors.

Many universities, community colleges, and technical schools have Continuing Education or Corporate Training Divisions that specialize in providing training for business and industry. Managers should know the administrator responsible for continuing education at all colleges, universities, or technical schools in their communities.

Partnerships with institutions of higher education offer several advantages to organizations that want to arrange training for their employees. Representatives of these institutions are education and training professionals. They know how to transform training objectives into customized curricula, courses, and lessons. They know how to deliver instruction and have access to a wide range of instructional support systems (libraries, multimedia centers, and instructional design centers). They know how to design application activities that simulate real-world conditions. They know how to develop a valid and reliable system of evaluation and use the results produced to chart progress and prescribe remedial activities when necessary.

In addition to professional know-how, institutions of higher education have resources that can markedly reduce the cost of training for an organization. Tuition costs for continuing education activities are typically much less than those associated with traditional college courses. If these institutions do not have faculty members on staff who are qualified to provide instruction in a given area, they can usually hire a temporary or part-time instructor who is qualified. Other advantages institutions of higher education can offer are credibility, formalization, standardization, and flexibility in training locations. Associating with a community college, university, or technical school can formalize an organization's training program and give it credibility.

This is important because employers sometimes find their attempts at customized training hampered by a lack of credibility. Their employees have been conditioned to expect formal grade reports, transcripts, and certificates of completion. These things formalize training in the minds of employees and make it more real for them. Educational institutions can provide these credibility builders.

EVALUATING TRAINING

Did the training provided satisfy the training objectives? Are trainees using what they learned? Has the training brought results? Managers need to know the answers to these questions every time training is provided. However, these can be difficult questions to answer. Evaluating training begins with a clear statement of purpose. What is the purpose of the training? This broad purpose should not be confused with the more specific training objectives. The purpose of the training is a broader concept. The objectives translate this purpose into more specific, measurable terms.

The purpose of training is to improve the knowledge, skills, and attitudes of employees and, in turn, the overall quality and productivity of the organization so that it becomes more competitive. In other words, the purpose of training is to improve performance and, in turn, competitiveness. To know whether training has improved performance, managers need to know three things:

- Was the training provided valid?
- Did the employees learn?
- Has the learning made a difference?

Valid training is training that is consistent with the training objectives. Evaluating training for validity is a two-step process. The first step involves comparing the written documentation for the training (course outline, lesson plans, curriculum framework, etc.) with the training objectives. If the training is valid in design and content, the written documentation will match the training objectives. The second step involves determining whether the actual training provided is consistent with the documentation. Training that strays from the approved plan will not be valid. Student evaluations of instruction conducted immediately after completion can provide information on consistency and the quality of instruction. Figure 8–3 is an example of an instrument that allows students to evaluate instruction.

Determining whether employees have learned is a matter of building evaluation into the training. Employees can be tested to determine whether they have learned, but be sure that tests are based on the training objectives. If the training is valid and employees have learned, the training should make a difference in their performance. Performance on the job should improve. This means quality and productivity should improve. Managers can make determinations about performance using the same indicators that told them training was needed in the first place. "Can employees perform tasks they could not perform before the training? Is waste reduced? Has quality improved? Is setup time down? Is in-process time down? Is the on-time rate up? Is the production rate up? Is throughput time down?" These are the types of questions managers should ask to determine whether training has improved performance.

Gilda Dangot-Simpkin of Dynamic Development suggests a checklist of questions for evaluating training programs that are purchased:[6]

- Does the program have specific behavioral objectives?
- Is there a logical sequence for the program?
- Is the training relevant for the trainee?

Instructions

On a scale of 1 to 5 (5 = highest rating; 1 = lowest rating), rank your instructor on each item. Leave blank any item which does not apply.

Organization of Course

1. Objectives (Clear to Unclear) 1 2 3 4 5
2. Requirements (Challenging to Unchallenging) 1 2 3 4 5
3. Assignments (Useful to Not Useful) 1 2 3 4 5
4. Materials (Excellent to Poor) 1 2 3 4 5
5. Testing Procedures (Effective to Ineffective) 1 2 3 4 5
6. Grading Practice (Explained to Not Explained) 1 2 3 4 5
7. Student Work Returned (Promptly to Delayed) 1 2 3 4 5
8. Overall Organization (Outstanding to Poor) 1 2 3 4 5

Comments

Teaching Skills

9. Class Meetings (Productive to Nonproductive) 1 2 3 4 5
10. Lectures (Effective to Ineffective) 1 2 3 4 5
11. Discussions (Balanced to Unbalanced) 1 2 3 4 5
12. Class Proceedings (To-the-Point/Wandering) 1 2 3 4 5
13. Provides Feedback (Beneficial to Not Beneficial) 1 2 3 4 5
14. Responds to Students (Positively/Negatively) 1 2 3 4 5
15. Provides Assistance (Always to Never) 1 2 3 4 5
16. Overall Rating of Instructor's Teaching Skills
    (Outstanding to Poor) 1 2 3 4 5

Comments

Substantive Value of Course

17. The course was (Intellectually Challenging to Elementary) 1 2 3 4 5
18. The instructor's command of the subject was
    (Broad and Accurate/Plainly Defective) 1 2 3 4 5
19. Overall substantive value of the course
    (Outstanding to Poor) 1 2 3 4 5

Comments

Figure 8–3
Student Evaluation of Instruction

■ Does the program allow trainees to apply the training?

■ Does the program accommodate different levels of expertise?

■ Does the training include activities that appeal to a variety of learning styles?

■ Is the philosophy of the program consistent with that of the organization?

■ Is the trainer credible?

■ Does the program provide follow-up activities to maintain the training on the job?

Using these questions as a guide, managers can make intelligent choices when selecting commercially produced training objectives.

MANAGERS AS TRAINERS

To be good trainers, managers should have such characteristics as a thorough knowledge of the topics to be taught; a desire to teach; a positive, helpful, cooperative attitude; strong leadership abilities; a professional attitude and approach; and exemplary behavior that sets a positive example.[7]

In addition to having these characteristics, managers should be knowledgeable about the fundamental principles of learning and the four-step teaching method. The principles of learning summarize much of what is known about how people learn best. It is important to conduct training in accordance with these principles. The four-step teaching method is a basic approach to conducting training that has proven effective over many years of use.

Principles of Learning

The principles of learning summarize what is known and widely accepted about how people learn. Trainers can do a better job of facilitating learning if they understand the following principles:

- *People learn best when they are ready to learn.*   You cannot make employees learn anything. You can only make them want to learn. Therefore, time spent motivating employees to want to learn is time well spent. Before beginning instruction, explain why employees need to learn and how they and the organization will mutually benefit from their having done so.

- *People learn more easily when what they are learning can be related to something they already know.*   Build today's learning on what was learned yesterday and tomorrow's learning on what was learned today. Begin each new learning activity with a brief review of the one that preceded it. Use examples that employees can all relate to.

- *People learn best in a step-by-step manner.*   This is an extension of the preceding principle. Learning should be organized into logically sequenced steps that proceed from the concrete to the abstract, from the simple to the complex, and from the known to the unknown.

- *People learn by doing.*   This is probably the most important principle for trainers to understand. Inexperienced trainers tend to confuse talking (lecturing or demonstrating) with teaching. These things can be part of the teaching process, but they do little good unless they are followed with application activities that require the learner to do something. Consider the example of teaching an employee how to ride a bicycle. You might present a thorough lecture on the principles of pedaling and steering and give a comprehensive demonstration on how to do it. However, until the employee gets on and begins pedaling, he or she will not learn how to ride a bicycle.

- *The more often people use what they are learning, the better they will remember and understand it.*   How many things have you learned in your life that you can no longer remember? People forget what they do not use. Trainers should keep this principle in mind. It means that repetition and application should be built into the learning process.

- *Success in learning tends to stimulate additional learning.*   This principle is a restatement of a fundamental principle in management (success breeds success). Organize training into long enough segments to allow learners to see progress but not so long that they become bored.

- *People need immediate and continual feedback to know if they have learned.*   Did you ever take a test and get the results back a week later? If so, that was probably a week later than you wanted them. People who are learning want to know immediately and continually how they are doing. Trainers should keep this principle in mind at all times. Feedback can be as simple as a nod, a pat on the back, or a comment such as "Good job!" It can also be more formal, such as a progress report or a graded activity. Regardless of the form it takes, trainers should concentrate on giving immediate and continual feedback.

Four-Step Teaching Method

Regardless of the setting, teaching is a matter of helping people learn. One of the most effective approaches for facilitating learning is not new, innovative, gimmicky, or high-tech in nature. It is known as the four-step teaching method, an effective approach to use for training. The four steps and a brief description of each follow:

- *Preparation* encompasses all tasks necessary to get participants prepared to learn, trainers prepared to teach, and facilities prepared to accommodate the process. Preparing participants means motivating them to want to learn. Personal preparation involves planning lessons and preparing all of the necessary instructional materials. Preparing the facility involves arranging the room for both function and comfort, checking all equipment to ensure it works properly, and making sure that all tools and other training aids are in place.

- *Presentation* is a matter of presenting the material participants are to learn. It might involve giving a demonstration, presenting a lecture, conducting a question/answer session, helping participants interact with a computer or interactive videodisc system, or assisting those who are proceeding through self-paced materials.

- *Application* is a matter of giving learners opportunities to use what they are learning. Application might range from simulation activities in which learners role-play to actual hands-on activities in which learners use their new skills in a live format.

- *Evaluation* is a matter of determining the extent to which learning has taken place. In a training setting, evaluation does not need to be a complicated process. If the training objectives were written in measurable, observable terms, evaluation is simple. Employees were supposed to learn how to do X, Y, and Z well and safely. Have them do X, Y, and Z and observe the results. In other words, have employees demonstrate proficiency in performing a task and observe the results.

WHY TRAINING SOMETIMES FAILS

Training is an essential ingredient in total quality, but training is not automatically good. In fact, training often fails. Training fails when it does for several reasons, such as poor teaching, inadequate curriculum materials, poor planning, insufficient funding, and a lack of commitment.

Some subtle and more serious reasons for training failures are explained by Juran as follows:[8]

- *Lack of participation in planning by management.* It is important to involve people at the line level in the planning of training. However, this does not mean management should be excluded. In fact, quite the opposite is true. Management must be involved, or the training may become task or technique oriented as opposed to results oriented. It is critical that training be results oriented, or in the long run it will fail.

- *Too narrow in scope.* Training that is to improve quality should proceed from the broad and general to the more specific. Often organizations jump right into the finite aspects of total quality such as statistical process control, just-in-time manufacturing, or teamwork before employees understand the big picture and where these finite aspects fit into it.

Writing for *Training*, Linda Harold makes the point that training sometimes fails because it focuses too specifically on how to more effectively and efficiently perform a task or complete a process instead of focusing on helping employees become independent thinking, creative problem solvers.[9] According to Harold:

> Over the past few years, many organizations have launched training programs for quality improvement. These programs are designed to teach employees specific processes that will improve quality. And, initially, significant improvements in quality do occur. Once these processes have been implemented, however, many organizations reach a plateau in quality improvement. Why? Employees have been trained in a process. They have not been allowed to think for themselves. When employees complete the process, they don't know how to take the next step because they haven't learned to think through the next step on their own.[10]

QUALITY TRAINING CURRICULUM FOR MANAGERS

For managers to play a leadership role in a total quality setting, they must be well trained in at least what Juran calls the Juran Trilogy: quality planning, quality control, and quality improvement.[11] A curriculum outline for each of these areas is provided in the following sections.

Quality Planning Training

Quality planning is the first component of the Juran Trilogy. According to Juran, quality planning should cover the following topics:[12]

- Strategic management for quality
- Quality policies and their deployment
- Strategic quality goals and their deployment
- The Juran Trilogy
- Big Q and little Q
- The triple-role concept
- Quality planning road map
- Internal and external customers
- How to identify customers
- Planning macroprocesses
- Planning microprocesses
- Product design
- Planning for process control
- Transfer to operations
- Santayana review (lessons learned)
- Planning tools

Quality Control Training

Quality control is the second component of the Juran Trilogy. According to Juran, quality control training should cover the following topics:[13]

- Strategic management for quality
- The feedback loop in quality control
- Controllability (self-control)
- Planning for control

- Control subjects
- Responsibility for control
- How to evaluate performance
- Interpretation of statistical and economic data for significance
- Decision making
- Corrective action
- Quality assurance audits
- Control tools

Quality Improvement Training

Quality improvement is the third component of the Juran Trilogy. According to Juran, quality improvement training should cover the following topics:[14]

- Strategic management for quality
- The Juran Trilogy
- Quality Council and its responsibilities
- Cost of poor quality: how to estimate it
- Project-by-project concept
- Estimating return on investment
- Nominating, screening, and selecting projects
- Infrastructure for quality improvement
- Macroprocess improvement projects
- Diagnostic journey
- Remedial journey
- Progress review
- Using recognition and reward to motivate
- Quality improvement tools and techniques

By standardizing the curriculum as set forth here and ensuring that all managers complete training in these three broad areas of quality, companies can come closer to achieving consistency of performance, and consistency of performance is critical in a total quality setting because it makes performance easier to measure and improve.

ENDNOTES

1. John Hoerr, "Business Shares the Blame for Workers' Low Skills," *Business Week* (25 June 1990): 71.
2. Tom Peters, *Thriving on Chaos: Handbook for a Management Revolution* (New York: Harper & Row, 1987), 386.
3. Nell P. Eurich, *Corporate Classrooms: The Learning Business* (Lawrenceville, NJ: Princeton University Press, 1985), x, xi.
4. Carolyn Wilson, *Training for Non-Trainers* (New York: AMACOM, 1990), 18–19.
5. Eurich, 16.
6. Gilda Dangot-Simpkin, "How to Get What You Pay For," *Training* (July 1990): 53–54.
7. This section is based on David L. Goetsch, *Industrial Safety and Health in the Age of High Technology* (Upper Saddle River, NJ: Prentice Hall/Merrill, 1993), 391–402.

8. Joseph M. Juran, *Juran on Leadership for Quality: An Executive Handbook* (New York: Free Press, 1989), 342.

9. Linda Harold, "The Power of LEARNING at Johnsonville Foods," *Training* (April 1991): 56.

10. Harold, 56.

11. Juran, 337.

12. Juran, 325.

13. Juran, 336.

14. Juran, 337.

CHAPTER NINE

Quality Tools

- Total Quality Tools Defined
- The Pareto Chart
- Cause-and-Effect Diagrams
- Check Sheets
- Histograms
- Scatter Diagrams
- Run Charts and Control Charts
- Stratification
- Some Other Tools Introduced
- Management's Role in Tool Deployment

One of the basic tenets of total quality is *management by facts*. This is not in harmony with the capability so revered in North America and the West in general: the ability to make snap decisions and come up with quick solutions to problems in the absence of input beyond intuition, gut feel, and experience. Management by facts requires that each decision, each solution to a problem, is based on relevant data and appropriate analysis. Once we get beyond the very small business (in which the data are always resident in the few heads involved, anyway), most decision points and problems will have many impacting factors, and the problem's root cause or the best-course decision will remain obscure until valid data are studied and analyzed. Collecting and analyzing data can be difficult. The total quality tools presented in this chapter make that task easy enough for anyone. Their use will assure better decision making, better solutions to problems, and even improvement of productivity and products and services.

Writing about the use of statistical methods in Japan, Dr. Kaoru Ishikawa said:

> The above are the so-called seven indispensable tools . . . that are being used by everyone: company presidents, company directors, middle management, foremen, and line workers. These tools are also used in a variety of [departments], not only in the manufacturing [department] but also in the [departments] of planning, design, marketing, purchasing, and technology.[1]

No matter where you fit into your organization today, you can use some or all of these tools to advantage, and they will serve you well for your future prospects.

This chapter explains the most widely used total quality tools and their applications, provides some insights on the involvement of management and the cross-functional nature of the tools, and issues some cautions.

TOTAL QUALITY TOOLS DEFINED

Carpenters use a kit of tools designed for very specific functions. Their hammers, for example, are used for the driving of nails. Their saws for the cutting of wood. These and others enable a carpenter to build houses. They are *physical* tools. Total quality tools also enable today's employees, whether engineers, technologists, production workers, managers, or office staff, to do their jobs. Virtually no one can function in an organization that has embraced total quality without some or all of these tools. Unlike those in the carpenter's kit, these are *intellectual* tools: they are not wood and steel to be used with muscle; they are tools for collecting and displaying information in ways to help the human brain grasp thoughts and ideas. When thoughts and ideas are applied to physical processes, the processes yield better results. When applied to problem solving or decision making, better solutions and decisions are developed.

The seven tools discussed in the following seven sections of this chapter represent those generally accepted as the basic total quality tools. We also discuss some others briefly later in this chapter. A case can be made that *just-in-time* and *statistical process control* are total quality tools. But these are more than tools: they are complete systems under the total quality umbrella. This book devotes an entire chapter to each of these systems.

A tool, like a hammer, exists to help do a job. If the job includes *continuous improvement*, *problem solving*, or *decision making*, then these seven tools fit the definition. Each of these tools is some form of chart for the collection and display of specific kinds of data. Through the collection and display facility, the data become useful information—information that can be used to solve problems, enhance decision making, keep track of work being done, even predict future performance and problems. The beauty of the charts is that they organize data so that we can immediately comprehend the message. This would be all but impossible without the charts, given the mountains of data flooding today's workplace.

THE PARETO CHART

The *Pareto* (pah-raý-toe) *chart* is a very useful tool wherever one needs to separate the important from the trivial. The chart, first promoted by Dr. Joseph Juran, is named after Italian economist/sociologist Vilfredo Pareto (1848–1923). He had the insight to recognize that in the real world a minority of causes lead to the majority of problems. This is known as the Pareto principle. Pick a category, and the Pareto principle will usually hold. For example, in a factory you will find that of all the kinds of problems you can name, only about 20% of them will produce 80% of the product defects; 80% of the cost associated with the defects will be assignable to only about 20% of the total number of defect types occurring.[2] Examining the elements of this cost will reveal that once again 80% of the total defect costs will spring from only about 20% of the cost elements.

Charts have shown that approximately 20% of the pros on the tennis tour reap 80% of the prize money and that 80% of the money supporting churches in the United States comes from 20% of the church membership.

All of us have limited resources. That point applies to you and to me, and to all enterprises—even to giant corporations and to the government. That means that our resources (time, energy, and money) need to be applied where they will do the most good. The purpose of the Pareto chart is to show you where to apply your resources by revealing the significant few from the trivial many. It helps us establish priorities.

The Pareto chart in Figure 9–1 labels a company's customers A, B, C, D, E, and All Others. The bars represent the percentage of the company's sales going to the respective customers. Seventy-five percent of this company's sales are the result of just two customers. If one adds customer C, 90% of its sales are accounted for. All the other customers together account for only 10% of the company's sales. Bear in mind that "Other"

Figure 9–1
Pareto Chart: Percentage of
Total Sales by Customer

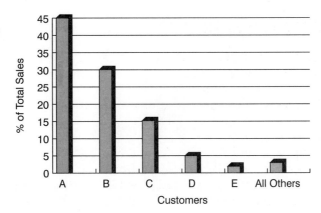

may include a very large number of small customers. Obviously, A, B, and perhaps C are the most critical customers. This would suggest that customers A, B, and C are the company's core market and all the others represent a marginal business. Decisions on where to allocate resources should be made accordingly.

The Pareto chart in Figure 9–2 shows bars representing the sales of a particular model of automobile by age group of the buyers. The curve represents the cumulative percentage of sales and is keyed to the y-axis scale on the right. Concentrating advertising on the 26–45 age bracket will result in the best return on investment because 76% of the Swift V-12 buyers come from the combination of the 36–45 and 26–35 age groups. The significant few referred to in the Pareto principle are in the 26–45 group. The insignificant many are all those under 26 and over 45.

Cascading Pareto Charts

Pareto charts may be cascaded by determining the most significant category in the first chart, then making a second chart related only to that category, and repeating this as far as possible, to three, four, or even five or more charts. If the cascading is done properly, root causes of problems may be determined rather easily.

Consider the following example. A company produces complex electronic assemblies, and the test department is concerned about the cost of rework resulting from test failures. They are spending more than $190,000 per year, and that amount is coming directly from profit. The department formed a special project team to find the cause of the problem and reduce the cost of rework. The Pareto chart in Figure 9–3 showed them that about 80% of the cost was related to just five defect causes. All the others, and there were about 30 more, were insignificant—*at least at that time.*

Figure 9–2
Swift V-12 Sales by Age Group

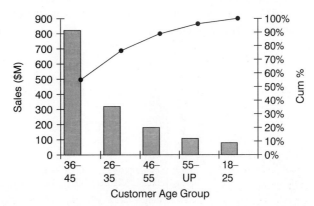

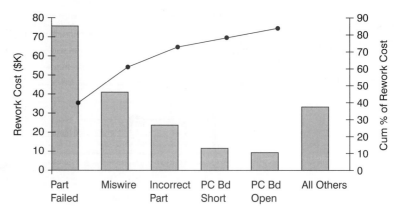

Figure 9–3
Top Five Defects by Rework Cost

The longest bar alone accounted for nearly 40% of the cost. If the problem it represented could be solved, the result would be an immediate reduction of almost $75,000 in rework cost. The team sorted the data again to develop a level 2 Pareto chart, Figure 9–4, to focus on any part types that might be a major contributor to the failures.

Figure 9–4 clearly showed that one type of relay accounted for about 60% of the failures. No other part failures came close. In this case, and at this time, the relay was the significant one, and all the other parts were the insignificant many. At this point, another team was formed to analyze the failure modes of the relay in order to determine a course of action for eliminating the relay problem. It was determined that there were a number of failure modes in the relay. They were plotted on the Pareto chart shown in Figure 9–5, which immediately revealed that 66% of all the failures were associated with one failure mode. The second longest bar in Figure 9–5 represented another manifestation of the same root cause. The relay contacts were not switching on *at all* (longest bar) or were not switching on *completely* (next longest bar). With this information known, the relay contacts were carefully examined, and it was determined that the relays were being damaged at incoming inspection where they were tested with a voltage that was high enough to damage the gold plating on the contacts. Changing the incoming test procedure and working with the relay vendor to improve its plating process eliminated the problem.

Earlier, we implied that although a particular problem might be insignificant at one point in time, it might not remain so. Consider what happens to the bars on the cascaded charts when the relay contact problem is solved. The second longest bar on the first chart clearly becomes the longest (assuming it was not being solved simultaneously with the relay problem). At this point, more than $100,000 a year is still being spent from profit to rework product rather than make it properly in the first place. The cycle must continue to be repeated until perfection is approached.

The next cycle of Pareto charts might look like those in Figure 9–6. Starting at the top we see the following points:

Figure 9–4
Rework Cost by Top Five Part
Failure Categories

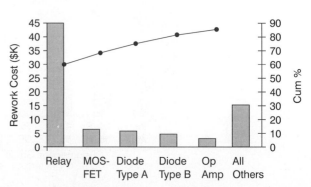

Figure 9–5
Relay Failure Categories

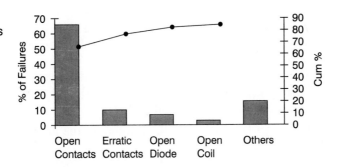

Figure 9–6
Second Cascading of Pareto Charts

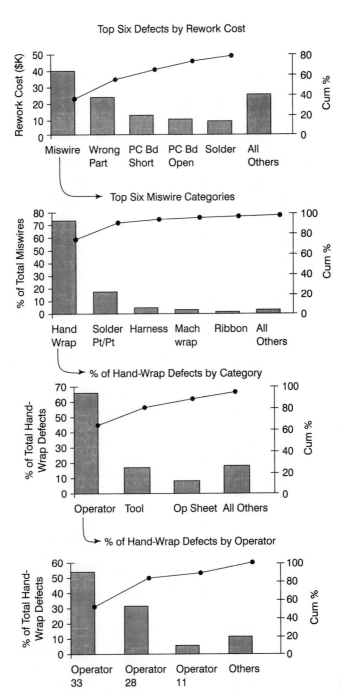

1. Miswires (wires connected to the wrong point or not properly attached to the right point) account for 40% of the remaining rework cost. (1st chart of Figure 9–6)
2. Wires connected with hand-wrapping tools represent more than 70% of all miswires. (2nd chart)
3. Of the hand-wrap defects, more than 65% are because of operator error. (3rd chart)
4. Of all the operators doing hand-wrap work, operators 33 and 28 contribute more than 80% of the defects. (4th chart)

Attention must be given to those operators in the form of training or, perhaps, reassignment.

A third Pareto chart cascading would break down the Wrong Part problem. For example, perhaps part abc is mistakenly substituted for part xyz on a printed circuit board. The cascading cycle may be repeated over and over, each time dealing with the significant few, while ignoring the trivial many. Eventually perfection is approached. A few companies are getting close with some of their products, but most have ample opportunity for significant improvement. One need not worry about running out of improvement possibilities.

CAUSE-AND-EFFECT DIAGRAMS

A team typically uses a *cause-and-effect diagram* to identify and isolate causes of a problem. The technique was developed by the late Dr. Kaoru Ishikawa, a noted Japanese quality expert, so sometimes the diagram is called an *Ishikawa diagram*. It is also often called a fishbone diagram because that is what it looks like.

In his book *Guide to Quality Control*, Ishikawa explains the benefits of using cause-and-effect diagrams as follows:

*   Creating the diagram itself is an enlightened, instructive process.
*   Such diagrams focus a group, thereby reducing irrelevant discussion.
*   Such diagrams separate causes from symptoms and force the issue of data collection.
*   Such diagrams can be used with any problem.[3]

The cause-and-effect diagram is the only tool of the seven tools that is not based on statistics. This chart is simply a means of visualizing how the various factors associated with a process affect the process's output. The same data could be tabulated in a list, but the human mind would have a much more difficult time trying to associate the factors with each other and with the total outcome of the process under investigation. The cause-and-effect diagram provides a graphic view of the entire process that is easily interpreted by the brain. Suppose an electronics plant is experiencing soldering rejects on printed circuit (PC) boards. People at the plant decide to analyze the process to see what can be done; they begin by calling together a group of people to get their thoughts. The group is made up of engineers, solder machine operators, inspectors, buyers, production control specialists, and others. All the groups in the plant who have anything at all to do with PC boards are represented. This is necessary to get the broadest possible view of the factors that might affect the process output.

The group is told that the issue to be discussed is the solder defect rate, and the objective is to list all the factors in the process that could possibly have an impact on the defect rate. The group uses brainstorming to generate the list of possible *causes*. The list might look like Figure 9–7.

The group developed a fairly comprehensive list of factors in the PC board manufacturing process—factors that could *cause* the *effect* of solder defects. Unfortunately, the list does nothing in terms of suggesting which of the 35 factors might be major causes, which are minor, and how they relate to each other. This is where the cause-and-effect diagram comes into play. Ishikawa's genius was to develop a means by which these random ideas might be organized to show relationships and to help people make intelligent choices.

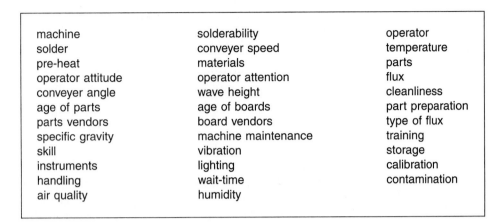

machine	solderability	operator
solder	conveyer speed	temperature
pre-heat	materials	parts
operator attitude	operator attention	flux
conveyer angle	wave height	cleanliness
age of parts	age of boards	part preparation
parts vendors	board vendors	type of flux
specific gravity	machine maintenance	training
skill	vibration	storage
instruments	lighting	calibration
handling	wait-time	contamination
air quality	humidity	

Figure 9–7
Brainstormed List of Possible Causes for Solder Defects

Figure 9–8 is a basic cause-and-effect diagram. The spine points to the *effect*. The effect is the "problem" we are interested in—in this case, machine soldering defects. Each of the ribs represents a cause leading to the effect. The ribs are normally assigned to the causes considered to be *major factors*. The lower-level factors affecting the major factors branch off the ribs. Examine Figure 9–7 to see whether the major causes can be identified. The ribs are assigned to these causes.

Six major groupings of causes are discernible:

1. The solder machine itself is a major factor in the process.
2. The operators who prepare the boards and run the solder machine would also be major factors.
3. The list includes many items such as parts, solder, flux, boards, and so forth, and these can be collected under the word *materials*, which also appears on the list. Materials is a major factor.
4. Temperature within the machine, conveyor speed and angle, solder wave height, and so on, are really the *methods* (usually published procedures and instructions) used in the process. Methods is a major factor.
5. Many of these same items are subject to the plant's methods (how-to-do-it) and measurement (accuracy of control), so measurement is a major factor, even though it did not appear on the list.
6. Cleanliness, lighting, temperature and humidity, and the quality of the air we breathe can significantly affect our performance and thus the quality of output of processes with which we work. We will call this major factor *environment*.

The designated six major factors, or causes, are those that the group thinks might have an impact on the quality of output of the machine soldering process: machine,

Figure 9–8
Basic Cause-and-Effect or
Fishbone Diagram

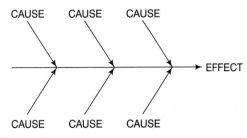

Figure 9–9
Cause-and-Effect Diagram with
Major Causes and Effect As-
signed

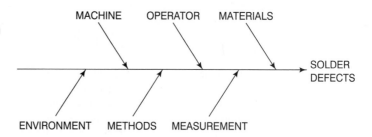

operator, materials, methods, measurement, and environment. The cause-and-effect
fishbone diagram developed from this information has six ribs as shown in Figure 9–9.

Having assigned the major causes, the next step is to assign all the other causes to
the ribs they affect. For example, *machine maintenance* should be assigned to the Machine rib, because machine performance is obviously affected by how well or how poorly
the machine is maintained. *Training* is attached to the Operator rib, because the degree to which operators have been trained certainly affects their expertise in running
the machine. In some cases, a possible cause noted on the list may appropriately branch
not from the rib (major cause) but from one of the branches (contributing cause). For
example, *solderability* (the relative ease—or difficulty—with which materials can be
soldered) would branch from the Materials rib, because it is a contributor to the materials' cause of solder defects. An important cause of poor solderability is age of parts.
So *age of parts* will branch not from Materials, but from solderability. Study Figure
9–10 to get a graphic sense of the relationships described in this paragraph.

Figure 9–10 is the completed fishbone diagram. It presents a picture of the major
factors that can cause solder defects and in turn the smaller factors that affect the major factors. Examination of the Materials rib shows that there are four factors directly
affecting materials as regards solder defects: the parts themselves, handling of the materials, and the solder and flux used in the process. The chart points out that contam-
ination can affect the solder's performance and also that the big issue affecting the parts
is solderability. In this case, the branches go to three levels from the rib, noting that
solderability can be affected by the vendor supplying the parts, storage of the parts before use, and age of the parts.

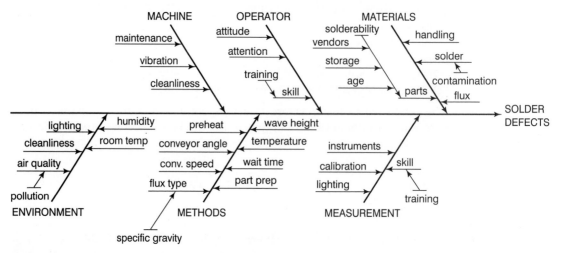

Figure 9–10
Completed Cause-and-Effect Diagram

Now you may say, "The diagram didn't configure itself in this way. Someone had to know the relationships before the diagram was drawn, so why is the diagram needed?" First, picture these relationships in your mind—no diagram, just a mental image. If you are not familiar with the process used in the example, pick any process involving more than two or three people and some equipment, such as the process of an athletic event. If you try this, you will probably find it virtually impossible to be conscious of all the factors coming into play, to say nothing of how they relate and interact. Certainly, the necessary knowledge and information already existed before the 35 factors were arranged in the cause-and-effect diagram. The key to the diagram's usefulness is that it is very possible that no *one* individual had all that knowledge and information. That is why cause-and-effect diagrams are normally created by teams of people widely divergent in their expertise.

The initial effort by the team is developing the list of possible factors. This is usually done using brainstorming techniques. Such a list can be made in a surprisingly short time—usually no more than an hour. It is not necessary that the list be complete or even that all the factors listed be truly germane. Missing elements will usually be obvious as the diagram is developed, and superfluous elements will become obvious and discarded. After the list has been assembled, all the team members contribute from their personal knowledge and expertise to assemble the cause-and-effect diagram.

The completed diagram reveals factors or relationships that had previously not been obvious. The causes most likely responsible for the problem (solder defects) will usually be isolated. Further, the diagram may suggest possibilities for action. It is possible in the example that the team, because it is familiar with the plant's operation, could say with some assurance that solderability was suspected because the parts were stored for long periods of time. They might recommend that, by switching to a just-in-time system, both storage and aging could be eliminated as factors affecting solderability.

The cause-and-effect diagram serves as an excellent reminder that the items noted on it are the things the company needs to pay attention to if the process is to continually improve. Even in processes that are working well, continuous improvement is the most important job any employee or team can have. In today's competitive global marketplace, it is truly the key to survival.

CHECK SHEETS

The *check sheet* is introduced here as the third of the seven traditional tools. The fuel that powers the total quality tools is data. In many companies, elaborate systems of people, machines, and procedures exist for the sole purpose of collecting data. At times, this quest for data has become zealous to the point of obscuring the reason for data collection in the first place. Many organizations are literally drowning in their own data while at the same time not knowing what is actually going on; they are "data-rich and information-poor." With the advent of powerful desktop computers, information collection has become an end unto itself in many instances.

Having access to data is essential. However, problems arise when trivial data cannot be winnowed from the important and when there is so much of it that it cannot be easily translated into useful information. Check sheets help deal with this problem.

The check sheet can be a valuable tool in a wide variety of applications. Its utility is restricted only by the imagination of the person seeking information. The check sheet can take any form. The only rules are that data collection must be the equivalent of entering a check mark and that the displayed data be easily translated into useful information. For example, it may take the form of a drawing of a product with the check marks entered at appropriate places on the drawing to illustrate the location and type of finish blemishes. An accounts receivable department might set up a check sheet to record the types and number of mistakes on invoices prepared. Check sheets apply to any work environment—not just to the factory floor.

The purpose of the check sheet is to make it easy to collect data for specific purposes and to present it in a way that facilitates conversion from data to useful information. For example, suppose we are manufacturing parts that have a specified dimensional tolerance of 1.120″ to 1.130″. During the week, each part is measured and the data is recorded. Figure 9–11 is a summary of the week's results.

This figure contains all the data on shaft length for the week of July 11. Without a lot of additional work, it will be difficult to glean much useful information from this list of data. Imagine how much more difficult it would be if, instead of a table, you were presented with a stack of computer runs several inches thick. That is frequently the case in the information age. (The information age should be called the *data age*, in our opinion, reflecting the difference between an abundance of raw, often meaningless data, and the real paucity of *useful* information.)

The computer could be programmed to do something with this data to make it more useful and in some situations that would be appropriate. After all, computers are good at digesting raw data and formatting it for human consumption. But before the computer can do that, some human must tell it exactly what it must do, how to format the information, what to discard, what to use, and so on. If we can't first figure out what to do with the data, no amount of computer power will help. On the supposition that we do know what to do with the data, it is possible that we could *preformat* the data so that it will be instantly useful *as it is being collected*. This is one of the powerful capabilities of the check sheet.

The importance of the data in Figure 9–11 rests in reporting how the work being produced relates to the shaft length specification. The machine has been set up to produce shafts in the center of the range so that normal variation would not spill outside the specified limits of 1.120″ and 1.130″ and thereby create waste. If the raw data could give us a feel for this as it is being collected, that would be very helpful. We would also like to know when the limits are exceeded. The check sheet in Figure 9–12 has been designed to facilitate both data collection and conversion to information.

The check sheet of Figure 9–12 is set up to accept the data very easily and at the same time display useful information. The check sheet actually produces a histogram

Figure 9–11
Weekly Summary of Shaft Dimensional Tolerance Results
(This is *not* a check sheet.)

Shaft length - Week of   7/11   (Spec: 1.120 - 1.130″)

Date	Length	Date	Length	Date	Length	Rem
11	1.124	11	1.128	11	1.123	
11	1.126	11	1.128	11	1.125	
11	1.119	11	1.123	11	1.122	
11	1.120	11	1.122	11	1.123	
12	1.124	12	1.126	12	1.125	
12	1.125	12	1.127	12	1.125	
12	1.121	12	1.124	12	1.125	
12	1.126	12	1.124	12	1.127	
13	1.123	13	1.125	13	1.121	
13	1.120	13	1.122	13	1.118	
13	1.124	13	1.123	13	1.125	
13	1.126	13	1.123	13	1.124	
14	1.125	14	1.127	14	1.124	
14	1.126	14	1.129	14	1.125	
14	1.126	14	1.123	14	1.124	
14	1.122	14	1.124	14	1.122	
15	1.124	15	1.121	15	1.123	
15	1.124	15	1.127	15	1.123	
15	1.124	15	1.122	15	1.122	
15	1.123	15	1.122	15	1.121	

Figure 9–12
Check Sheet of Shaft Dimensional Tolerance Results

Check Sheet

Shaft length - Week of ___7 /11___ (Spec: 1.120 - 1.130")

- -

1.118**   13

1.119**   11   ** Out of Limits

1.120   11 13

1.121   12 13 15 15

1.122   11 11 13 14 14 15 15 15

1.123   11 11 11 13 13 13 14 15 15 15

1.124   11 12 12 12 13 13 14 14 14 15 15 15

1.125   11 12 12 12 12 13 13 14 14

1.126   11 12 12 13 14 14

1.127   12 12 14 15

1.128   11 11

1.129   14   **Enter day of month for data point.**

1.130

1.131**

1.132**

- -

as the data are entered. (See the following section for information about histograms.) Data are taken by measuring the shafts, just as was done for Figure 9–11. But rather than logging the measured data by date, as in Figure 9–11, the check sheet in Figure 9–12 only requires noting the date (day of month) opposite the appropriate shaft dimension. The day-of-month notation serves as a check mark while at the same time keeping track of the day the reading was taken.

This check sheet should be set up on an easel on the shop floor, with entries handwritten. That will make the performance of the machine continuously visible to all—operators, supervisors, engineers, or anyone else in the work area.

The data in Figure 9–11 are the same as the data in Figure 9–12. Figure 9–11 shows columns of sterile data that, before meaning can be extracted, must be subjected to hard work at someone's desk. Assuming it does get translated into meaningful information, it will probably still remain invisible to the people who could make the best use of it—the operators. That can, of course, be overcome by more hard work, but in most cases the data will languish. On the other hand, Figure 9–12 provides a simple check sheet into which the data are entered more easily and once entered, provide a graphic presentation of performance. Should the check sheet reveal that the machine is creeping away from the center of the range, or if the histogram shape distorts, the operator can react immediately. No additional work is required to translate the data to useful information, and no additional work is required to broadcast the information to all who can use it.

To set up a check sheet, you must think about your objective. In this example, we were making shafts to a specification. We wanted to know how well the machine was performing, a graphic warning whenever the machine started to deviate, and information about defects. Setting the check sheet up as a histogram provided all the information needed. This is called a *Process Distribution Check Sheet* because it is concerned with the variability of a process. Other commonly used check sheets include Defective Item Check Sheets (detailing the variety of defects), Defect Location Check Sheets (showing where on the subject product defects occur), Defect Factor Check Sheets (illustrating the factors—time, temperature, machine, operator—possibly influencing defect generation), and many others.

If we wanted to better understand what factors might be contributing to excessive defects on the shop floor, we could set up a Defect Factors Check Sheet. As an example, go back to the section on Pareto charts and look at Figure 9–6. The top chart there revealed that miswires were the most significant defect in terms of cost. A check sheet could be set up to collect data about the factors that might be contributing to the miswire defects. We are primarily concerned with the operators themselves and the factors that may influence their performance, so the check sheet will list each operator's number and bench location within the factory. To determine whether the day of the week or the time of day has anything to do with performance, the data will be recorded by day and by morning or afternoon. We could have included tool numbers as well, but using a tool that produces faulty connections is something the operator must guard against. In other words, we will not consider a tool to be at fault—only the operator if he or she continues to use a defective tool.

In the check sheet, shown in Figure 9–13, five types of miswire defects, covering all types experienced, are coded by symbols, and these symbols are the only raw data entered on the chart. Sums of all defect categories are shown at the bottom of each column, and the weekly total for each operator is shown at the end of each row. A quick glance at the check sheet points to operators 28 and 33 as the sources of the problem. We don't know the *cause* at this point, but we know where to start looking.

In times past, these two people might very well have been summarily fired. In a total quality setting, that decision would be considered the last resort. Most people want

Operator No.	Bench No.	11/2 AM	11/2 PM	11/3 AM	11/3 PM	11/4 AM	11/4 PM	11/5 AM	11/5 PM	11/6 AM	11/6 PM	Week Totals
8	A3					●				o □		o - 1 / ● - 1 3 / □ - 1
10	A2			● ●			□					● - 2 / □ - 1 3
11	B1	o						o				o - 2 2
13	A1		o			△					□ □	o - 1 / △ - 1 4 / □ - 2
28	C2	o ● / o	o △	o o / o	△ o	o	o o	o △	□ o	o o	o ● / o	o - 17 / ● - 2 23 / △ - 3 / □ - 1
33	C3	o o / o △ / o	● o / o o	o △ / o o	o o / □	o ●	□ o / o	△ o / o	o o	o o / o o / △ o	o o / o o	o - 28 / ● - 2 36 / △ - 4 / □ - 2
40	B2	+							o ●			o - 1 / ● - 1 3 / + - 1

o = Hand Wrap   10 7 | 9 6 | 4 7 | 6 6 | 10 9 | 39 35
● = Solder PT/PT   17 | 15 | 11 | 12 | 19 | 74
△ = Harness
+ = Ribbon   □ = Other

Figure 9–13
Check Sheet: Defect Factors—Miswires

to do a good job and will if they are provided with the necessary resources and training. In a case like this, it is not unusual to find that the fault lies with management. The employees were not adequately trained for the job, or some environmental factor (noise, temperature, lighting, or something else) is at fault, or the operators may simply not be equipped for the task (because of vision impairment, impaired motor skills, or some other problem). In any of those scenarios, management is at fault and therefore should do the morally right thing to correct the problem.

Check sheets can be valuable tools for converting data into useful and easy-to-use information. The key is teaching operators how to use them and empowering them to do so.

HISTOGRAMS

Histograms are used to chart frequency of occurrence. How often does something happen? Any discussion of histograms must begin with an understanding of the two kinds of data commonly associated with processes: *attributes* and *variables* data. Although they were not introduced as such, both kinds of data have been used in the illustrations of this chapter. An *attribute* is something that the output product of the process either has or does not have. From one of the examples (Figure 9–6), an electronic assembly either had wiring errors or it did not. Another example (Figure 9–32) shows that an assembly either had broken screws or it did not. These are attributes. The example of making shafts of a specified length (Figures 9–11 and 9–12) was concerned with *variables data*. That example used shaft length measured in thousandths of an inch, but any scale of measurement can be used, as appropriate for the process under scrutiny. A process used in making electrical resistors would use the scale of electrical resistance in ohms, another process might use a weight scale, and so on. *Variables data* are something that results from measurement. See Figure 9–14.

Using the shaft example again, an all-too-common scenario in manufacturing plants would have been to place a Go-No Go screen at the end of the process, accepting all shafts between the specification limits of 1.120″ and 1.130″ and discarding the rest. Data might have been recorded to keep track of the number of shafts that had to be scrapped. Such a record might have looked like Figure 9–15, based on the original data.

Figure 9–15 tells us what we wanted to know if we were interested only in the number of shafts accepted versus the number rejected. Looking at the shaft process in this way, we are using *attributes data*: either they passed or they failed the screening. This reveals only that we are scrapping between 3% and 4% of all the shafts made. It does not reveal anything about the process adjustment that may be contributing to the scrap rate. Nor does it tell us anything about how robust the process is—might some

---

Attributes Data
- Has or has not
- Good or bad
- Pass or fail
- Accept or reject
- Conforming or nonconforming

Variables Data
- Measured values (dimension, weight, voltage, surface, etc.)

Figure 9–14
Characteristics of Attributes Data and Variables Data

Shaft Acceptance—Week of ___7/11___ (Spec: 1.120 - 1.130")		
Date	Accepted	Rejected
11.	11	1
12.	12	0
13.	11	1
14.	12	0
15.	12	0
Totals:	58	2

Figure 9–15
Summary Data: Weekly Shaft Acceptance

slight change push the process over the edge? For that kind of insight we need *variables data*.

One can gain much more information about a process when variables data are available. The check sheet of Figure 9–12 shows that both of the rejects (out of limits shafts) were on the low side of the specified tolerance. The peak of the histogram seems to occur between 1.123″ and 1.124″. If the machine was adjusted to bring the peak up to 1.125″, some of the low-end rejects might be eliminated without causing any new rejects at the top end. The frequency distribution also suggests that the process as it stands now will always have occasional rejects—probably in the 2%–3% range at best.

Potential Trap with Histograms

Be aware of a potential trap when using histograms. The histogram is nothing more than a measurement scale across one axis (usually the *x*-axis) and frequency of like measurements on the other. (Histograms are also called *frequency distribution diagrams*.) The trap occurs when measurements are taken over a long period of time. Too many things can affect processes over time: wear, maintenance, adjustment, material differences, operator influence, environmental influence. The histogram makes no allowance for any of these factors. It may be helpful to consider a histogram to be the equivalent of a snapshot of the process performance. If the subject of a photograph is moving, the photographer must use a fast shutter speed to prevent a blurred image. If the histogram data are not collected over a suitably short period of time, the result will be blurred, just as if the camera's shutter was too slow for the action taking place, because it is possible that the process's performance changes over time. Blurred photographs and blurred histograms are both useless. A good histogram will show a crisp snapshot of process performance as it was at the time the data were taken, not before and not after. This leads some people to claim that histograms should be used only on processes that are known to be *in control*. (See the section on control charts.)

That limitation is not necessary as long as you understand that histograms have this inherent flaw. Be careful that any interpretation you make has accounted for time and its effect on the process you are studying. For example, we do not know enough about the results of the shaft-making process from Figure 9–12 to predict with any certainty that it will do as well next week. We don't know that a machine operator didn't tweak the machine two or three times during the week, trying to find the center of the range. What happens if that operator is on vacation next week? Would we dare predict that performance will be the same? We can only make these predictions if we know the process is statistically in control, thus the warnings. Taking this into consideration, the histogram in Figure 9–12 still provides valuable information.

Histograms and Statistics

Understanding a few basic facts is fundamental to the use of statistical techniques for quality and process applications. We have said that all processes are subject to variability, or variation. There are many examples of this. One of the oldest and most graphically convincing is the bead experiment.[4] This involves a container with a large number of beads. The beads are identical except for color. Suppose there are 900 white beads and 100 red, making 1,000 total. The beads are mixed thoroughly (step 1). Then 50 beads are drawn at random as a sample (step 2). The red beads in the sample are counted. A check mark is entered in a histogram column for that number. All the beads are put back into the container, and they are mixed again (step 3). When you repeat these steps a second time, the odds are that a different number of red beads will be drawn. When a third sample is taken, it will probably contain yet another number of red beads. The process (steps 1, 2, and 3) has not changed, yet the output of the process does change. This is *process variation* or *variability*. If these steps are repeated over and over until a valid statistical sampling has been taken, the resulting histogram will invariably take on the characteristic bell shape common to process variability (see Figure 9–16).

It is possible to calculate the *process variability* from the data. The histogram in Figure 9–16 was created from 100 samples of 50 beads each. The data was as shown in Figure 9–17.

The flatter and wider the frequency distribution curve, the greater the process variability. The taller and narrower the curve, the less the process variability. Even though the variability may change from process to process, it would be helpful to have a common means of measuring, discussing, or understanding variability. Fortunately we do. To express the process's variability we need to know only two things, both of which can be derived from the process's own distribution data: *standard deviation* and *mean*. Standard deviation is represented by the lowercase Greek letter sigma (σ) and indicates a deviation from the average, or *mean*, value of the samples in the data set. The mean is represented by the Greek letter mu (μ). In a normal histogram, μ is seen as a vertical line from the peak of the bell curve to the base, and it is the line from which deviation is measured, minus to the left of μ and plus to the right. Standard deviation (σ) is normally plotted at -3σ, -2σ, -1σ (left of μ), and $+1\sigma$, $+2\sigma$, and $+3\sigma$ to the right (refer to Figure 9–19). Because mean and standard deviation are always derived from data from the process in question, standard deviation has a constant meaning from process to process. From this we can tell what the process can do in terms of its statistical variability (assuming that it remains stable, no changes introduced):

- 68.26% of all sample values will be found between $+1\sigma$ and -1σ.
- 95.46% of all sample values will be found between $+2\sigma$ and -2σ.
- 99.73% of all sample values will be found between $+3\sigma$ and -3σ.

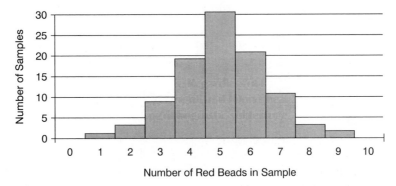

Figure 9–16
Frequency Distribution of Red Beads in Samples

Figure 9–17
Data on Red Beads in Samples

Samples with 0 red beads	0
Samples with 1 red bead	1
Samples with 2 red beads	3
Samples with 3 red beads	9
Samples with 4 red beads	19
Samples with 5 red beads	31
Samples with 6 red beads	21
Samples with 7 red beads	11
Samples with 8 red beads	3
Samples with 9 red beads	2
Samples with 10 red beads	0
Total samples taken	100

This information has a profound practical value, as we shall see as we develop the discussion.

Calculating the Mean

For a bell curve representing a truly normal distribution between $\pm$ infinity, the mean value would be a vertical line to the peak of the bell. Our curve is slightly off normal (because we are using a relatively small sample), so we must calculate μ:

$$\mu = \Sigma \chi \div n$$

We take the sum of all the χ values (number of samples $\times$ red beads in corresponding samples) and divide by the total number of samples (n) (see Figure 9–18):

$$510 \div 100 = 5.1$$

The μ is placed at 5.1 on the x-axis, and all deviations are measured relative to that. (See Figure 9–19.)

Calculating Standard Deviation

The formula for standard deviation:

$$\sigma = \sqrt{\Sigma d^2 / n - 1}$$

d = the deviation of any unit from the average
n = the number of units sampled

Red beads	# Samples	X value	d	d² ×	# Samples	=
0	0	0	−5.1	26.01	0	0
1	1	1	−4.1	16.81	1	16.81
2	3	6	−3.1	9.61	3	28.83
3	9	27	−2.1	4.41	9	39.69
4	19	76	−1.1	1.21	19	22.99
5	31	155	−0.1	0.01	31	0.31
6	21	126	.09	0.81	21	17.01
7	11	77	1.9	3.61	11	39.71
8	3	24	2.9	8.41	3	25.23
9	2	18	3.9	15.21	2	30.42
10	0	0	4.9	24.01	0	0
Totals	100 (n)	510 (Σx)			Σd²	= 221

Figure 9–18
Calculating Standard Deviation

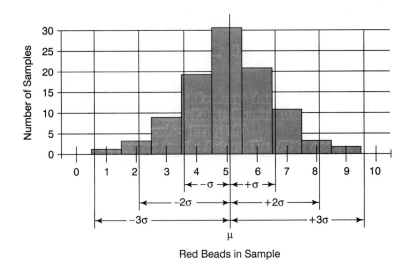

Figure 9–19
Application of Standard Deviation Calculations to Red Bead Histogram

To calculate the standard deviation, σ, square each of the deviations, d. The first deviation is the single sample containing one red bead. In Figure 9–18, it is in the one-bead row, which is at a distance of -4.1 from μ. $(-4.1^2 = 16.81)$. Next we have three samples at the two-bead row, at a distance of -3.1 from μ. $(-3.1^2 = 9.61$, but we have three samples in this row, so we count 28.83). This is repeated through row 10. Then we add all these to obtain Σd^2. Σd^2 is called the *sum of the squared deviations*. Next, divide by $n - 1 (100 - 1 = 99)$ or

$$221 \div 99 = 2.23$$

This is the *mean squared deviation*. The square root of $2.23 = 1.49$. This is the *root mean squared deviation* and is σ, or the standard deviation. (Note: calculations are to two decimal places.)

Next calculate the positions of $\mu \pm 1\sigma, 2\sigma$, and 3σ.

$$\sigma = 1.49 \qquad 2\sigma = 2.99 \qquad 3\sigma = 4.47$$

These values are entered on Figure 9–19:

$\mu - 1\sigma = 5.1 - 1.49 = 3.61$
$\mu + 1\sigma = 5.1 + 1.49 = 6.59$
$\mu - 2\sigma = 5.1 - 2.99 = 2.11$
$\mu + 2\sigma = 5.1 + 2.99 = 8.09$
$\mu - 3\sigma = 5.1 - 4.47 = 0.63$
$\mu + 3\sigma = 5.1 + 4.47 = 9.57$

Suppose we have a process that is operating like the curve in Figure 9–19. We have specifications for the product output that require us to reject any part below 3.6 and above 6.6. It turns out that these limits are approximately $\pm 1\sigma$. We know immediately that about one-third of the process output will be rejected. If this is not acceptable, which is highly probable, we will have to improve the process or change to a completely different process. Even if more variation could be tolerated in the product and we took the specification limits out to 2 and 8, about 5 of every 100 pieces flowing out of the process would still be rejected. In a competitive world, this is poor performance indeed. Many companies no longer consider 2,700 parts per million defective ($\pm 3\sigma$) to be good enough. Some are bold enough to suggest that six-sigma (3.4 parts per million) is achievable. Refer to Figure 9–20. Motorola has made significant progress toward achieving it

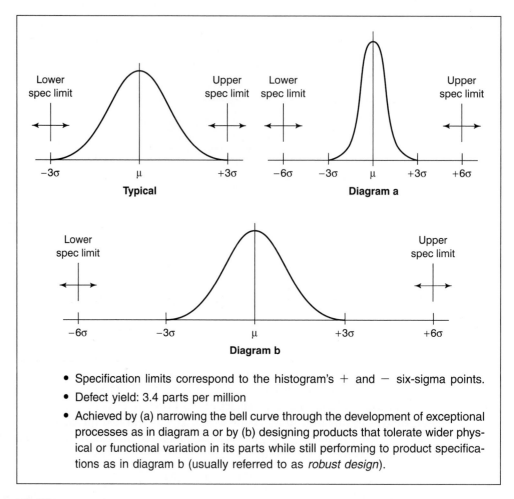

Figure 9–20
Six Sigma Process Capability

and others are following. Whatever the situation, with this statistical sampling tool properly applied, there is no question about what can be achieved with any process. You will be able to predict the results.

Shapes of Histograms

Consider the shape of some histograms and their position relative to specification limits. Figure 9–21 is a collection of histograms. Histogram A represents a normal distribution. So does B, except it is shallower. The difference between the process characteristics of these two histograms is that process A is much tighter, whereas the looser process B will have greater variances. Process A is usually preferred. Processes C and D are skewed left and right. Although the curves are normal, product will be lost because the processes are not centered. Process E is bimodal. This can result from two batches of input material, for example. One batch produces the left bell curve, and the second batch the curve on the right. The two curves may be separated for a better view of what is going on by stratifying the data by batch. (See the "Stratification" section.)

Histogram F suggests that someone is discarding the samples below and above a set of limits. This typically happens when there is a 100% inspection and only data that are within limits are recorded. The strange Histogram G might have used data from in-

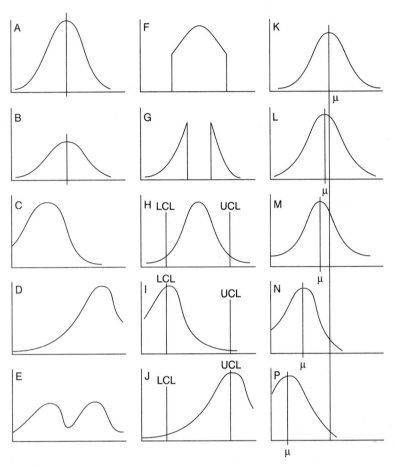

Figure 9–21
Histograms of Varying Shapes

coming inspection. The message here is that the vendor is screening the parts, and someone else is getting the best ones. A typical case might be electrical resistors which are graded as 1%, 5%, and 10% tolerance. The resistors that met 1% and 5% criteria were screened out and sold at a higher price. You got what was left.

Histogram H shows a normal distribution properly centered between a set of upper and lower control limits. Histograms I and J illustrate what happens when the same normal curve is allowed to shift left or right. There will be a significant loss of product as a control limit intercepts the curve higher up its slope.

Histograms K through P show a normal, centered curve that went out of control and drifted. Remember that histograms do not account for time and you must therefore be careful about making judgments. If all the data that produced Histograms K through P were averaged, or even if all the data were combined to make a single histogram, you could be misled. You would not know that the process was drifting. Plotting a series of histograms over time, such as K through P, clearly illustrates any drift right or left, shallowing of the bell, and the like.

The number of samples or data points has a bearing on the accuracy of the histogram, just as with other tools. But with the histogram there is another consideration: how does one determine the proper number of intervals for the chart? (The intervals are, in effect, the data columns of the histogram.) For example, Figure 9–16 is set up for 11 intervals: 0, 1, 2, and so on. The two outside intervals are not used, however, so the histogram plots data in nine intervals. The rule of thumb is as follows:

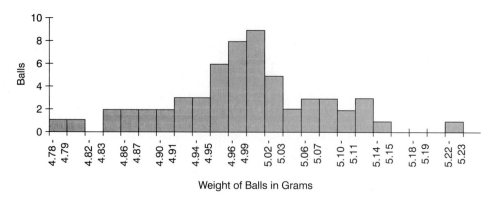

Figure 9–22
Histogram with Limited Amount of Data Stretched

Number of Observations (N)	Number of Intervals (k)
< 75	5–7
75–300	6–10
> 300	10–20

Or you may use the formula $k = \sqrt{N}$.

It is not necessary to be very precise with this. These methods are used to get close and adjust one way or the other for a fit with your data.

Suppose we are using steel balls in one of our products and the weight of the ball is critical. The specification is 5 ± 0.2 grams. The balls are purchased from a vendor, and because our tolerance is tighter than the vendor's, we weigh the balls and use only those that meet our specification. The vendor is trying to tighten its tolerance and has asked for assistance in the form of data. Today 60 balls were received and weighed. The data were plotted on a histogram. To give the vendor the complete information, a histogram with intervals every 0.02 grams is established.

Figure 9–22 does not look much like a bell curve because we have tried to stretch a limited amount of data (60 observations) too far. There are 23 active or skipped intervals. Our rule of thumb suggests 5 to 7 intervals for less than 75 observations. If the same data were plotted into a histogram of six intervals (excluding the blank), it would look like Figure 9–23. At least in this version it looks like a histogram. With more data—say, 100 or more observations—one could narrow the intervals and get more granularity. Don't try to stretch data too thin because the conversion to real information can become difficult and risky.

Figure 9–23
Histogram with Appropriate
Intervals for the Amount of Data

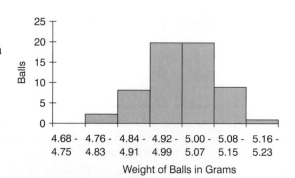

SCATTER DIAGRAMS

The fifth of the seven tools is the *scatter diagram*. It is the simplest of the seven and one of the most useful. The scatter diagram is used to determine the correlation (relationship) between two characteristics (variables). Suppose you have an idea that there is a relationship between automobile fuel consumption and the rate of speed at which people drive. To prove, or disprove, such an assumption, you could record data on a scatter diagram that has miles per gallon (mpg) on the *y*-axis and miles per hour (mph) on the *x*-axis; mpg and mph are the two characteristics.

Examination of the scatter diagram of Figure 9–24 shows that the aggregate of data points contains a slope down and to the right. This is correlation, and it supports the thesis that the faster cars travel, the more fuel they use. Had the slope been upward to the right, as it actually appears to be (for three of the four cars) between 20 and 30 mph, then the correlation would have suggested that the faster you travel, the better the fuel mileage. Suppose, however, that the data points did not form any recognizable linear or elliptical pattern but were simply in a disorganized configuration. This would suggest that there is no correlation between speed and fuel consumption.

Figure 9–25 is a collection of scatter diagrams illustrating strong *positive correlation* (Diagram A), weak *negative correlation* (Diagram B), and *no correlation* (Diagram C). To be classified as a strong correlation, the data points must be tightly grouped in a linear pattern. The more loosely grouped, the less correlation, and therefore the term *weak correlation*. When a pattern has no discernible linear component, it is said to show no correlation.

Scatter diagrams are useful in testing the correlation between process factors and characteristics of product flowing out of the process. Suppose you want to know whether conveyor speed has an effect on solder quality in a machine soldering process. You could set up a scatter diagram with conveyor speed on the *x*-axis and solder rejects or nonconformities on the *y*-axis. By plotting sample data as the conveyor speed is adjusted, you can construct a scatter diagram to tell whether a correlation exists.

In this case, Figure 9–26 suggests that the correlation is a curve, with rejects dropping off as speed is initially raised but then increasing again as the conveyor speed continues to increase. This is not atypical of process factors that have optimum operating points. In the case of the conveyor, moving too slowly allows excess heat to build up, causing defects. So increasing speed naturally produces better results, until the speed increases to the point where insufficient preheating increases the number of defects. Figure 9–26, then, not only reveals a correlation but also suggests that there is an optimum conveyor speed, operation above or below which will result in increased product defects.

It is also possible to determine a correlation between two process factors. If your manufacturing process includes the washing of parts in a cleaning agent, and you are

Figure 9–24
Scatter Diagram: Speed versus
Fuel Consumption for Four
Automobiles

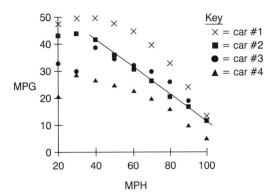

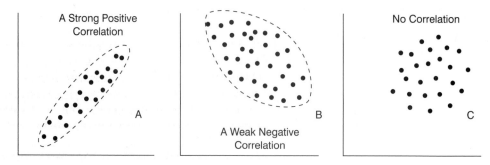

Figure 9–25
Scatter Diagrams of Various Correlations

interested in reducing the time the parts are in the cleaning tank, you might want to know whether the temperature of the solution is correlated with the time it takes to get the parts thoroughly clean. The scatter diagram could have temperature of the cleaning agent on one axis and time to clean on the other. By adjusting the temperature of the solution and plotting the cleaning time, a scatter diagram will reveal any existing correlation.

Assume that the scatter diagram shows a discernible slope downward to the right, as in Figure 9–27. This shows that over the temperature range tested, there *is* a correlation between cleaning solution temperature and cleaning time. With this information, you might be able to reduce the cycle time of the product. *Cycle time* in manufacturing is basically elapsed time from the start of your build process until the product is finished. Cycle time is becoming more important as manufacturers adopt world-class techniques to compete in the global marketplace. If you can find a safe, cost-effective way to raise the cleaning agent temperature to some more efficient level, and in the process shorten the cycle (or perhaps maintain the cycle and do a better job of cleaning), doing so might provide a competitive advantage.

Not all scatter diagrams require that special tests be run to acquire raw data. The data are frequently readily available in a computer. Few companies would have to record new data to determine whether a correlation exists between the day of the week and employee performance. Such data are often available from the day-to-day inspection reports. In fact, where people are involved, it is advisable to use existing data rather than collecting new data to be sure that the data were not influenced by the test itself. Imagine people being told they were to be part of a test to determine whether their performance was as good on Friday or Monday as the rest of the week, and then not reacting to that. This knowledge would undoubtedly affect their performance.

Figure 9–26
Scatter Diagram: Conveyor
Speed versus Rejects

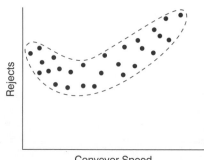

Figure 9–27
Scatter Diagram: Cleaning Solution versus Cleaning Time

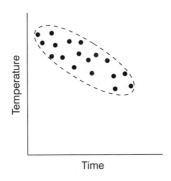

RUN CHARTS AND CONTROL CHARTS

The run chart is straightforward, and the control chart is a much more sophisticated outgrowth of it. Therefore, the two are usually thought of together as a single tool. Both can be very powerful and effective for the tracking and control of processes, and they are fundamental to the improvement of processes.

Run Charts

The *run chart* records the output results of a process over time. The concept is strikingly simple, and indeed it has been used throughout modern times to track performance of just about everything. Because one axis (usually the *x*-axis) represents time, the run chart can provide an easily understood picture of what is happening in a process as time goes by. That is, it will cause trends to "jump-out" at you. For this reason the run chart is also referred to as a *trend chart*.

Consider as an example a run chart set up to track the percentage or proportion of product that is defective for a process that makes ballpoint pens. These are inexpensive pens, so production costs must be held to a minimum. On the other hand, many competitors would like to capture our share of the market, so we must deliver pens that meet the expectations of our customers—as a minimum. A sampling system is set up that requires a percentage of the process output to be inspected. From each lot of 1,000 pens, 50 will be inspected. If more than 1 pen from each sample of 50 is found defective, the whole lot of 1,000 will be inspected. In addition to scrapping the defective pens, we will attempt to discover why the defects were there in the first place and to eliminate the cause. Data from the sample will be plotted on a run chart. Because we anticipate improvements to the process as a result of this effort, the run chart will be ideal to show whether we are succeeding.

The run chart of Figure 9–28 is the result of sample data for 21 working days. The graph clearly shows that significant improvement in pen quality was made during the 21 working days of the month. The trend across the month was toward better quality

Figure 9–28
Run Chart: Pen Defect Rate for 21 Working Days

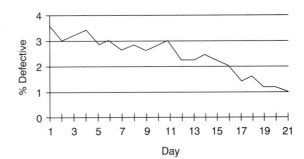

(fewer defects). The most significant improvements came at the twelfth day and the seventeenth day, as causes for defects were found and corrected.

The chart can be continued indefinitely to keep us aware of performance. Is it improving, staying the same, or losing ground? Scales may have to change for clarity. For example, if we consistently found all samples with defects below 2%, it would make sense to change the *y*-axis scale to 0%–2%. Longer-term charts would require changing from daily to weekly or even monthly plots.

Performance was improved during the first month of the pen manufacturing process. The chart shows positive results. What cannot be determined from the run chart, however, is what *should* be achieved. Assuming we can hold at 2 defective pens of 100, we still have 20,000 defective pens out of a million. Because we are sampling only 5% of the pens produced, we can assume that 19,000 of these find their way into the hands of customers—the very customers our competition wants to take away from us. So it is important to improve further. The run chart will help, but a more powerful tool is still needed.

Control Charts

The problem with the run chart and, in fact, many of the other tools, is that it does not help us understand whether the variation is the result of *special causes*—things such as changes in the materials used, machine problems, lack of employee training—or *common causes* that are purely random. Not until Dr. Walter Shewhart made that distinction in the 1920s was there a real chance of improving processes through the use of statistical techniques. Shewhart, then an employee of Bell Laboratories, developed the control chart to separate the *special causes* from the *common causes*.[5]

In evaluating problems and finding solutions for them, it is important to distinguish between special causes and common causes. Figure 9–29 shows a typical control chart. Data are plotted over time, just as with a run chart; the difference is that the data stay between the upper control limit (UCL) and the lower control limit (LCL) while varying about the center line or average—*only so long as the variation is the result of common causes (i.e., statistical variation)*. Whenever a special cause (nonstatistical cause) impacts the process, one of two things will happen: either a plot point will penetrate UCL or LCL, or there will be a "run" of several points in a row above or below the average line. When a penetration or a lengthy run appears, this is the control chart's signal that something is wrong that requires immediate attention.

As long as the plots stay between the limits and don't congregate on one side or the other of the process average line, the process is in statistical control. If either of these conditions is not met, then we can say that the process is not in statistical control or simply is "out of control"—hence the name of the chart.

It is the UCL, LCL, and process average lines added to the run chart that make the difference. The positioning of the lines cannot be arbitrary. Nor can they merely reflect what you want out of the process, for example, based on a specification. Such an approach won't help separate common causes from special causes, and it will only complicate at-

Figure 9–29
Basic Control Chart

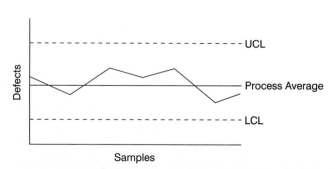

tempts at process improvement. UCL, LCL, and process average must be determined by valid statistical means.

All processes have built-in variability. A process that is in statistical control will still be affected by its natural random variability. Such a process will exhibit the normal distribution of the bell curve. The more finely tuned the process, the less deviation there will be from the process average, the narrower the bell curve. (Refer to Figure 9–21, Histogram A and Histogram B.) This is at the heart of the control chart and is what makes it possible to define the limits and process average.

Control charts are the appropriate tool to monitor processes. The properly used control chart will immediately alert the operator to any change in the process. The appropriate response to that alert is to stop the process at once, preventing the production of defective product. Only after the special cause of the problem has been identified and corrected should the process be restarted. Having eliminated a problem's root cause, that problem should never recur. (Anything less, however, and it is sure to return eventually.) Control charts also enable continuous improvement of processes. When a change is introduced to a process that is operated under statistical process control (SPC) charts, the effect of the change will be immediately seen. You know when you have made an improvement. You also know when the change is ineffective or even detrimental. This validates effective improvements, which you will retain. This is enormously difficult when the process is not in statistical control, because the process instability masks the results, good or bad, of any changes deliberately made.

To learn more about statistical process control and control charts, study Chapter 11. "Optimizing and Controlling Processes through Statistical Process Control (SPC)."

STRATIFICATION

Stratification is a simple tool in spite of its name. It involves investigating the cause of a problem by grouping data into categories. This grouping is called *stratification*. The groups might include data relative to the environment, the people involved, the machine(s) used in the process, materials, and so on. Grouping of data by common element or characteristic makes it easier to understand the data and to pull insights from it.

Consider an example from a factory floor. One of the factory's products requires five assemblers, all doing the same thing at the same rate. Their output flows together for inspection. Inspection has found an unacceptably high rate of defects in the products. Management forms a team to investigate the problem with the objective of finding the cause and correcting it. They plot the data taken over the last month (see Figure 9–30).

The chart in Figure 9–30 plots all operator-induced defects for the month. The team believes that for this product, zero defects can be approached. If you were going to react to this chart alone, how would you deal with the problem? You have five assemblers. Do they all contribute defects equally? This is hardly ever the case. The data can be stratified by the operator to determine each individual's defect performance. The charts in Figure 9–31 do this.

Figure 9–30
Chart of Operator Defects for November

Figure 9–31
Stratified Charts for Each Operator

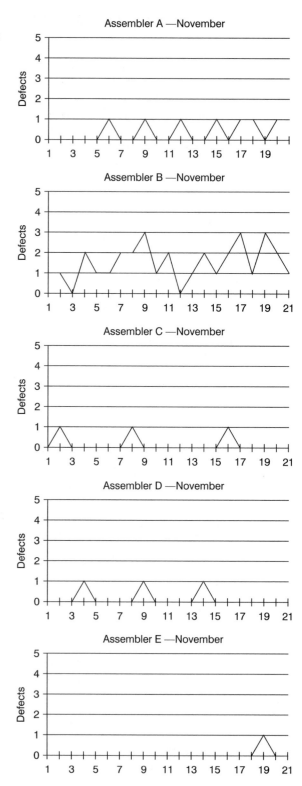

The five stratified charts in Figure 9–31 indicate that one operator, Assembler B, is responsible for more defects than the other four combined. Assembler A also makes more than twice as many errors as Assembler C or Assembler D and eight times as many as Assembler E, the best performer of the group.

The performance of Assembler A and Assembler B must be brought up to the level of the others. Possible causes of the operator-induced defects could be inherent skill, training, vision, attitude, attentiveness, and environmental factors such as noise, lighting, and temperature in the operator's work station area. The charts provide an indication of the place to start making changes.

The Pareto charts of Figure 9–6 on page 119 also represent *stratification*. Figure 9–6 started with a series of defect types that were the most costly (the first chart). Then it took the worst case, Miswires, and divided it into the *kinds* of miswires (the second chart). Then the worst kind, Hand-Wrap, was split into several categories (the third chart). The dominant Hand-Wrap defect category was operator induced. Finally, the Operator category was stratified by individual operator (the fourth chart).

The power of stratification lies in the fact that if you stratify far enough you will arrive at a *root cause* of the problem. Only when root causes are corrected will the problem be solved. Any other kind of solution is a *work-around* fix. Work-arounds are often used in the real world, but when they are, the underlying problem remains and will eventually cause disruption again.

In the present example, we probably did not go all the way to the root cause, unless Assembler B has serious mental, vision, or motor problems that could not be corrected. The most likely root cause is that Assembler B has not been adequately trained for the job, something readily ascertained when the focus is on that individual. One or two more charts looking at the time of day when the mistakes are being made might yield some information, but once the problem is isolated to a person, discussion will usually take you quickly to the root cause. If, on the other hand, Assembler B is a robot and not a human (which is entirely possible in today's automated environment), the stratification should go to at least one more level. We would have to determine the kinds of defects that Assembler B (the robot) is making. That may lead to an adjustment or repair of the machine.

Figure 9–32 shows that the category of defects induced by this machine are almost all concerned with screws. The robot is either damaging the screws or breaking them off. Show this chart to the robot maintenance technician, and that person will immediately recognize that the robot needs an adjustment or replacement of its torque controller. The root cause of the problem is either misadjustment or a defective controller. The technician can confirm the diagnosis by running tests on the robot before certifying it for return to service.

Data collected for Pareto charts and run charts (Figure 9–30) can be stratified. Virtually any data can be subjected to stratification. This includes the data collected for control charts, check sheets, histograms, and scatter diagrams. Consider an example of a stratified scatter diagram.

Scatter diagrams, which show the relationship between the x- and y-axis, lend themselves well to stratification. In this example parts are being finished on two identical machines. A scatter diagram is plotted to correlate surface flatness and machine speed.

Figure 9–33 suggests that there is a correlation between machine speed (revolutions per minute, or rpm) and surface flatness between 500 and 1,000 rpm but no

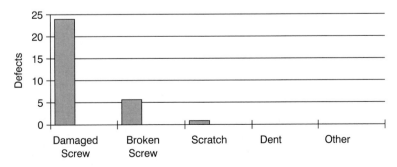

Figure 9–32
Robot B Defect Category for November

Figure 9–33
Scatter Diagram: Surface Flatness versus Revolutions per Minute

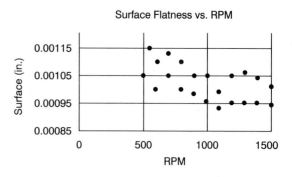

correlation at higher revolutions per minute. When the same data are *stratified* in the charts of Figure 9–34, the picture becomes clearer.

In Figure 9–34, the charts reveal that the two machines react similarly to speed increase, but Machine 1 is better than Machine 2 by about 0.0001" in its ability to produce a flat surface. The Machine 1 chart also suggests that increases beyond 1,000 rpm do not produce much improvement. A finish of 0.00095 is about as good as the machine will produce. On the other hand, the Machine 2 chart does show some improvement (two data points) past 1,300 rpm. Given the difference between the two machines, one message coming from the charts is that Machine 2 should be examined to determine the cause of its poorer performance. (More than likely it will be found that bearing wear is the factor in question and that can be corrected easily.) After the machine has been repaired, new data should be taken to verify that 1,000–1,100 rpm is the best practical machine speed.

The charts in Figure 9–34 indicate another message. Both machines had data points better than normal at 550 and 1,100 rpm. It appears that the machines have a natural resonance that affects performance. The clue here is that both machines show it at 550 rpm and at double that speed (1,100 rpm). This should be checked out, because it could be adversely affecting performance across the range. If vibration and resonance could be "quieted" across the operating range as it apparently is at 550 and 1100 RPM, then the performance might be significantly improved in both machines. The data that gave us this signal are in the scatter diagram of Figure 9–33, but they don't jump out at you the way they do in the stratified charts of Figure 9–34.

In these examples, we have stratified assembly defects by operator, machine-induced defects by type of defect, and machine performance by machine. It was also pointed out that the earlier Pareto chart discussion involved stratification in which defects were stratified to types of defects, the worst of which was in turn stratified to the processes

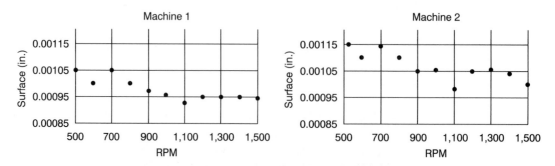

Figure 9–34
Stratified Scatter Diagrams: Surface Flatness versus Revolutions per Minute

producing those defects. The process (Hand-Wrap) producing the most defects was stratified to process factors, and finally the factor revealed as the most significant (Operator) was stratified to individual operators.

There is virtually no limit on the directions stratification can take. For example, the operators could have been stratified by age, training, sex, marital status, by teams, experience, or other factors. The machines could have been stratified by age, date of maintenance, tools, and location (and in the case of similar but not identical machines, by make and model number). In similar fashion, operating procedures, environment, inspection, time, materials, and so forth, can be introduced.

SOME OTHER TOOLS INTRODUCED

The preceding sections have discussed the statistical tools that have come to be known as the Seven Tools. One should not conclude, however, that these seven are the only tools needed for pursuing world-class performance. These seven are the ones that have been found most useful for the broadest spectrum of users. Ishikawa refers to them as the "seven indispensable tools for quality control."[6] He goes on to say that they are being used by everyone from company presidents to line workers and across all kinds of work—not just manufacturing. These seven probably represent the seven basic methods most useful to all the people in the workplace. We recommend three more as necessary to complete the tool kit of any business enterprise, if not each of the players within the business:

- The flow diagram
- The survey
- Design of experiments (DOX)

Both Deming[7] and Juran[8] promote the use of flow diagrams. Ishikawa includes surveys and design of experiments in his *Intermediate Statistical Method* and *Advanced Statistical Method*,[9] respectively.

Flowcharts

A *flowchart* is a graphic representation of a process. A necessary step in improving a process is to flowchart it. In this way, all parties involved can begin with the same understanding of the process. It may be revealing to start the flowcharting process by asking several different team members who know the process to flowchart it independently. If their charts are not the same, one problem is revealed at the outset. Another strategy is to ask team members to chart how the process *actually* works and then chart how they think it *should* work. Comparing the two versions can be an effective way to identify causes of problems and to suggest improvement possibilities. The most commonly used flowcharting method is to have the team, which is made up of the people who work within the process and those who provide input to or take output from the process, work together to develop the chart. To be effective, the completed flowchart must accurately reflect the *way the process actually works,* not how it should work. After a process has been flowcharted, it can be studied to determine what aspects of it are problematic and where improvements can be made.

You may already be familiar with the flowchart, at least to the point of recognizing one when you see it. It has been in use for many years and in many ways. The application we have in mind here is for flowcharting the inputs, steps, functions, and outflows of a process to more fully understand how the process works, who or what has input to and influence on the process, its inputs and outputs, and even its timing.

A set of standard flowcharting symbols for communicating various actions, inputs, outflows, and so forth, are used internationally. They may be universally applied to any

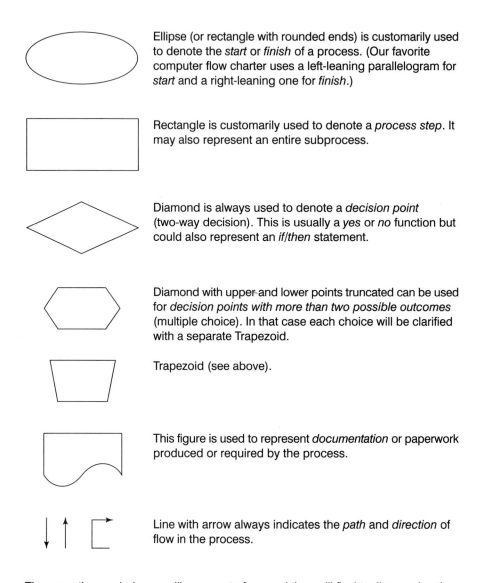

Ellipse (or rectangle with rounded ends) is customarily used to denote the *start* or *finish* of a process. (Our favorite computer flow charter uses a left-leaning parallelogram for *start* and a right-leaning one for *finish*.)

Rectangle is customarily used to denote a *process step*. It may also represent an entire subprocess.

Diamond is always used to denote a *decision point* (two-way decision). This is usually a *yes* or *no* function but could also represent an *if/then* statement.

Diamond with upper-and lower points truncated can be used for *decision points with more than two possible outcomes* (multiple choice). In that case each choice will be clarified with a separate Trapezoid.

Trapezoid (see above).

This figure is used to represent *documentation* or paperwork produced or required by the process.

Line with arrow always indicates the *path* and *direction* of flow in the process.

These are the symbols you will use most often, and they will fit virtually any situation.

Table 9–1
Standard Symbology for Flowcharts

process. The most commonly used symbols are shown in Table 9–1. To illustrate their use, a simple flow diagram using the most common symbol elements is given in Figure 9–35. Flow diagrams may be as simple or as complex as you may need. For example, in Figure 9–35 the rectangle labeled "Troubleshoot" represents an entire subprocess that itself can be expanded into a complex flowchart. If an intent of the flowchart had been to provide information on the troubleshooting process, then each troubleshooting step would have to be included. Our purpose for Figure 9–35 was merely to chart the *major* process steps for receiving and repairing a defective unit from a customer, so we did not require subprocess detail. This is a common starting point. From this high-level flowchart, it may be observed that the customer's defective unit is (a) received, (b) the problem is located and corrected, and (c) the repaired unit is tested. (d) If the unit fails the test, it is recycled through the repair process until it does pass. (e) Upon passing the test, paperwork is completed. (f) Following that, the customer is notified, and (g) the unit is returned to the customer along with a bill for services. With this high-level flowchart as a guide, your next step will be to develop detailed flowcharts

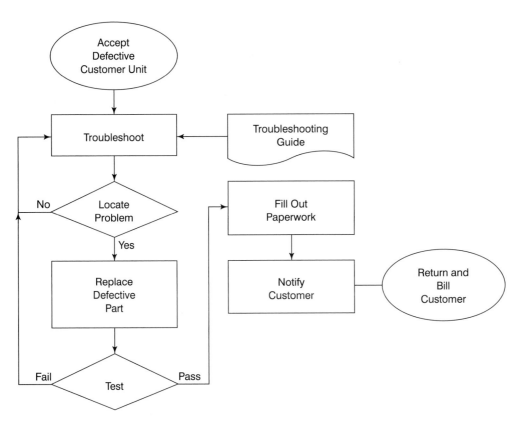

Figure 9–35
Typical Processes Flow Diagram

of the subprocesses you want to improve. Only then can you understand what is really happening inside the process, which steps add value and which do not, where the time is being consumed, identify redundancies, and so on. Once you have a process flow-charted, it is almost always easy to see potential for improvement and streamlining. Without the flowchart it may be impossible.

More often than not, people who work directly with a process are amazed to find out how little understanding of their process they had before it had been flowcharted. Working with any process day in and day out tends to breed a false sense of familiarity.

Processes tend to grow and become more complex and less efficient over time. This can occur without being noticed by the individuals involved in the process. A process flowchart can readily reveal such situations.

If you set out to control or improve any process, it is essential that you fully understand the process and why it is what it is. Don't make the assumption that you already know or that the people working in the process know, because chances are good that you don't, and they don't. Work with the people who are directly involved, and flowchart the process as a first step in the journey to world-class performance. Not only will you better understand how the processes work, but you will spot unnecessary functions or weaknesses and be able to establish logical points in the process for control chart application. Use of the other tools will be suggested by the flowchart as well.

Surveys

All of the tools are designed to present information—information that is pertinent, easily understood by all, and valuable for anyone attempting to improve a process or enhance the performance of some work function. The purpose of a survey is to obtain relevant information from sources that otherwise would not be heard from—at least not in the context of providing helpful data. Because you design your own survey, you can tailor it to your needs. We believe that the survey meets the test of being a total quality tool. Experience has shown that the survey can be very useful.

Surveys can be conducted internally, as a kind of employee feedback on problem areas, or as *internal customer* feedback on products or services. They can also be conducted with *extenal customers,* your business customers, to gain information about how your products or services rate in the customers' eyes. The customer (internal or external) orientation of the survey is important, because the customer, after all is said and done, is the only authority on the quality of your goods and services. Some companies conduct annual customer satisfaction surveys. These firms use the input from customers to focus their improvement efforts.

Surveys are increasingly being used with suppliers as well. We are finally coming to the realization that having a huge supplier base is not the good thing we thought it was. The tendency today is to cut back drastically on the number of suppliers utilized, retaining those that offer the best *value* (not best price, which is meaningless) and that are willing to enter into partnership arrangements. If a company goes this route, it had better know how satisfied the suppliers are with the past and present working relationship and what they think of future prospects. The survey is one tool for determining this. It is possibly the best initial method for starting a supplier reduction/supplier partnership program.

Even if you are not planning to eliminate suppliers, it is vital to know what your suppliers are doing. It would make little sense for you to go to the trouble of implementing total quality if your suppliers continue to do business as usual. As you improve your processes and services and products, you cannot afford to be hamstrung by poor quality from your suppliers. Surveys are the least expensive way of determining where suppliers stand on total quality and what their plans are for the future. The survey can also be a not-too-subtle message to the suppliers that they had better get on the bandwagon.

A typical department in any organization has both internal suppliers and internal customers. Using the same customer-oriented point of view in a survey has proven to be a powerful tool for opening communications among departments and getting them to work together for the common goal, rather than for department glory—usually at the expense of the overall company.

The downside of surveys is that the right questions have to be asked, and asked in ways that are unambiguous and designed for short answers. A survey questionnaire should be thoroughly thought out and tested before it is put into use. Remember that you will be imposing on the respondents' time, so make it easy and keep it simple.

Design of Experiments

Design of experiments (DOX) is a very sophisticated method for experimenting with processes with the objective of optimizing them. If you deal with complicated processes that have multiple factors affecting them, DOX may be the only practical way of bringing about improvement. Such a process might be found in a wave soldering machine, for example. Wave solder process factors include these:

Solder type	Conveyor speed	Flux specific gravity
Solder temperature	Conveyor angle	Wave height
Preheat temperature	PC board layer count	Flux type
PC board groundplane mass		

These 10 factors influence the process, often interacting with one another. The traditional way to determine the proper selection/setting was to vary one factor while holding all others fixed. That kind of experimentation led to making hundreds of individual runs for even the simplest processes. With that approach it is unusual to arrive at the optimum setup, because a change in one factor frequently requires adjustment of one or more of the other factors for best results.

Design of experiments reduces the number of runs from hundreds to tens as a rule, or by an order of magnitude. This means process experimentation allows multiple factor adjustment simultaneously, shortening the total process, but equally as important, revealing complex interaction among the factors. A well-designed experiment can be concluded on a process such as wave soldering in 30–40 runs and will establish the optimum setting for each of the adjustable parameters for each of the selected factors. For example, optimal settings for conveyor speed, conveyor angle, wave height, preheat temperature, solder temperature, and flux specific gravity will be established for each PC board type, solder alloy, and so on.

DOX will also show which factors are critical and which are not. This information will enable you to set up control charts for those factors that matter, while saving the effort that might have been expended on the ones that don't. While design of experiments is beyond the scope and intent of this book, the DOX work of Deming, Taguchi, and others may be of help to you. Remember that DOX is available as a tool when you start trying to improve a complex process.

MANAGEMENT'S ROLE IN TOOL DEPLOYMENT

Management's role is changing from one of directing to one of facilitating. Since the Industrial Revolution, management has supplied the place of work, the machinery and tools, and the work instructions. The concept has been that management knows what the job is and needs only to hire the muscle power to get it accomplished. The workers were there only because management could not get the job done without their labor. Workers were not expected to think about doing things differently but simply to follow the boss's orders. Work was typically divided into small tasks that required minimal training, with little or no understanding on the part of laborers as to how their contribution fit into the mosaic of the whole.

During much of the 20th century, and certainly after World War II, changes began creeping into the management–labor relationship. Some people think that the labor unions were responsible for these changes, and they did help obtain better pay, shorter hours, workplace improvements, and other things for workers. However, the relationship changes between management and labor have happened largely in spite of the unions. Unions have had at least as difficult a time as management has had in dealing with employee involvement. Nor has management at large been responsible for the changes sweeping across the industrial world today. Certainly there are champions representing management, but the changes are coming about for one reason and one reason only: they are necessary in order for businesses to survive in a marketplace that has been made increasingly competitive by the Japanese who, alone, followed the quality teachings of Deming and Juran (and expanded on them through the work of Ishikawa, Taguchi, Shingo, Ohno, and others).

Survival mentality finally surfaced in the United States in the 1980s. We woke up to the fact that not only our industrial survival but perhaps even our national survival was at stake. We had to become more competitive in the global marketplace.

Now that the wake-up call has been received, many people have come to realize that we have been managing poorly for a very long time—say, since 1945. We (those of us who have heard the alarm) have come to understand that management's proper role is to facilitate, not to direct. Management provides the place of work and the machines and tools as before, but in addition, we must do everything we can to *help* our employees do the job. That means training. It means listening to their thoughts and ideas—more than that—*seeking* their thoughts and ideas. It means acting on them. It means giving them the power to do their jobs without management interference. It means giving them time to think and discuss and suggest and experiment. It means communicating—fully and honestly. No secrets, no smoke screens. It means accepting every employee as a valued member of the corporate team.

This approach does not mean that management abdicates its responsibility to set the direction for the enterprise, to establish the corporate vision, to steer the course. But with the enlistment of all the brain power that had formerly gone untapped, even this job becomes easier than it was before.

It is management's responsibility to train employees to use not only physical tools (and that is very important) but also intellectual tools. The seven tools discussed in this chapter should eventually be used by every employee—*eventually* because it is a mistake to schedule all employees for training on the tools if they will not be using them very soon. When a group of people are ready to put some of the tools to practice, that is when they should be trained. As the total quality concept takes root, it will be only a matter of time until everyone has the need. Train them as required.

Management must also provide the internal experts, often called facilitators, to help the new teams get started and to develop their expertise. Facilitation is probably a neverending function, because the total quality envelope is constantly being expanded, and there will always be the need for a few to be on the leading edge, to bring the others along.

It is management's responsibility to ensure that the people who are solving the problems have the proper training and facilitation. It is also management's responsibility to make sure the problems being attacked are of interest to the enterprise and not trivial. Management must populate the problem-solving team with the cross-functional expertise the problem requires. The team must be given the power and support necessary to see the effort brought to its conclusion.

Management must be vigilant that data used in problem solving are valid, which is a function that usually falls to the facilitator. Especially when teams are immature in total quality, they have a tendency to grab at the first set of data that comes along. It is management's responsibility to ensure that the data and the statistical techniques employed are appropriate for the problem at hand.

Finally, management must ensure that there are results. Too many problem-solving, process improvement, and related efforts take on a life of their own, and go on forever. This cannot be allowed. People are watching. Especially in the early stages, some people will hold the view, referring to total quality, that "This too shall pass." If results do not come rather quickly, the detractors will be given the ammunition they need to subvert the whole effort. For this reason, it is important that the first projects attempted have a high probability of success, and management must monitor them closely, even to the point of being involved in the activity. As the process matures and successes are tallied, an occasional failure will not be an issue. In fact, people must be given the chance to fail, and failure must be free of repercussions for the team or its members.

Precautions

Implementing the use of statistical tools and the whole concept of process improvement, problem solving by the rank and file, empowerment—in short, the total quality culture—represents a profound change from the way things have been done in the past. People generally resist change until they see that it will benefit them. For that reason, management must become the champions of change and convince everyone that the effort will benefit all. Those who would undermine the effort must rapidly be converted or removed from the operation. People will be looking to management for evidence that management really believes in total quality. If for no other reason than that, it must be obvious to all that management is using the same techniques the other employees are being taught. Above all, management must support and facilitate the employees as they use the techniques of total quality to solve problems and improve processes.

Although results should be evident very soon, do not expect the necessary cultural change to occur overnight. This is a long process, requiring several years to get to the point where total quality is considered "just the way we do things" and not some special "project." Even so, during all that time, problems are being solved, improvements made, and efficiency, productivity and competitiveness improved.

■ *Communicate.*   Let everyone know what is going on and what the results are. Help them understand why it is good for them, the whole enterprise, and, yes, even for the nation.

■ *Never assume that you know it all.*   The people who live with the processes day in and day out know far more about what is wrong with them and how to improve them than any manager. Never delude yourself that you have learned all you need to know about total quality. It will never happen because it is a dynamic and ever-expanding concept.

■ *Start slowly.*   Don't try to organize an entire factory or office complex into improvement teams and train everyone in sight on day one. Take it one or two steps at a time, training as you go. Be careful to pick early projects that have high prospects for success.

■ *But start.*   The worst choice a manager could make today is to decide that total quality is not for his or her business. It is for every conceivable kind of business, large or small, whether public, private, military, civilian, mass production, job shop, classroom, or office. It would be a tragedy to decide not to start this journey when so much is at stake.

ENDNOTES

1. Kaoru Ishikawa, *Guide to Quality Control* (Tokyo: Asian Productivity Organization, 1976).
2. The 80–20 rule is an approximation, and one should not expect the numbers to land exactly at 80% or 20%.
3. Ishikawa, *Guide to Quality Control*, 24–26.
4. Joseph M. Juran, *Juran on Planning for Quality* (New York: Free Press, 1988), 180.
5. W. W. Scherkenbach, *The Deming Route to Quality and Productivity* (Rockville, MD: Mercury, 1991), 100.
6. Ishikawa, *What Is Total Quality Control? The Japnese Way* (Englewood Cliffs, NJ: Prentice Hall, 1985), 198.
7. Scherkenbach, 104.
8. Juran, 18.
9. Ishikawa, *What Is Total Quality Control?* 199.

Problem Solving and Decision Making

Problem solving and decision making are fundamental to total quality. On the one hand, good decisions will decrease the number of problems that occur. On the other hand, the workplace will never be completely problem-free. The purpose of this chapter is threefold:

- Learning to solve problems effectively, positively, and in ways that don't create additional problems
- Becoming better decision makers
- Learning to make decisions and handle problems in ways that promote quality

PROBLEM SOLVING FOR TOTAL QUALITY

If you ask the typical manager to describe his or her biggest problem in today's workplace, the response will probably include one or more of the following:

- We spend all our time in meetings trying to resolve problems.
- We are constantly fighting problems, and that doesn't leave us time to do our real jobs, such as planning, leading, and so forth.
- As soon as we put out one fire, another pops up.
- We've got more problems than we can handle, and it bogs us down.

The actual words may vary, but the message is the same. The workplace can be so burdened with problems that managers and others spend so much time trying to fix them that nothing gets done right. Leadership suffers—there is just no time to lead. Performance suffers, from the standpoint of both the individual and the organization. Quality of product or service deteriorates. Competitiveness is negatively impacted. Failure of the organization becomes a real probability, especially if its competitors have turned to total quality and its philosophy for solving and eliminating problems—once and for all. Why is it that with all the effort we put into it, consuming so much time in the process, we cannot solve our problems and get on with the jobs we are paid to do? The answer is simply that most of our problem solving does not solve problems.

Consider this. We were once driving along in our car and the engine quit. Turn the key and it fired up; release the key and it quit again. We found that there was a resistor in the electrical system that was only used after the engine had been started, and it had failed, hence no electrical energy to the spark plugs. By replacing the resistor with a simple piece of wire the engine would run fine. We put the wire in place and drove home. Had we solved the problem? We suspected that we had not, because if a penny's worth of copper wire could have been used in place of a part that cost a couple of dollars, then surely Chrysler would have opted for the wire. We suspected that our fix was just a Band-Aid. Sure enough, the next day the ignition coil failed. That was a $10 part, and it failed because by substituting the wire for the defective resistor we had put too much voltage on the coil, causing it to self-destruct. We had not understood the reason for the resistor being there. We did not have all the data we should have had before using that piece of wire. The lessons here: (a) Get data before you try to solve any problem, or you may make things even worse. (b) Band-Aid fixes do not solve problems and may cause new, unforeseen problems.

At this point we spent $12 for a replacement resistor and an ignition coil (note that that was six times what we would have spent if we had properly replaced the resistor in the first place). The car ran fine. Problem solved? Most people would say so.

This is the level of problem solving in most organizations. When something breaks, fix it or replace it. Job done, problem solved. The most that we should claim for this type of problem solving is that we are back where we started (i.e., before the problem came up). But remember, if it happened once it can happen again.

Getting back to the Chrysler, the resistor failed twice more while we owned the car. And over the ensuing years there were two more Chrysler products in our family, both of which had multiple failures of the same resistor. Replacing the resistor did not solve the problem. For the problem to be truly solved so that the part would last for the car's expected life would have required Chrysler to gather all the electrical, physical, and reliability data relating to the resistor, and the circuit it operated in, and then redesign the circuit or make use of a more robust resistor or some other change, as the data required. Had Chrysler done that, we could justifiably call it a solved problem. We could also call such a solution a product *improvement*, because the probability of failure would be greatly reduced. (We have had no firsthand experience with Chrysler cars for several years, so they may have indeed solved it. We hope so.)

The point we want to make here is that in total quality jargon, a problem is only solved when it is rendered impossible or significantly less probable to recur. That will always be the objective of total quality problem solving. Any problem that is merely fixed by restoring the situation to what it was before the problem was manifested will return again. That is why our managers spend so much time with problem issues. The problems are not being solved, just put into a recycle loop. In those organizations that have adopted total quality, problems are solved once and for all. The same problems do not return time and time again. That means that there will be fewer problems tomorrow than there were today, fewer next month than this month, fewer next year than this year. Managers will have more time to manage, leaders to lead. With problem solutions leading to process or products/service improvement,

- product or service quality improves,
- costs decrease (through less waste and warranty action),
- customer satisfaction improves,
- competitiveness improves, and
- the probability for success improves.

Clearly all of these outcomes are desirable. And they are all achievable by applying the total quality principles to problem solving.

SOLVING AND PREVENTING PROBLEMS

Even the best-managed organizations have problems. A problem is any situation in which what exists does not match what is desired. Said another way, a problem yields a discrepancy between the current and the desired state of affairs. The greater the disparity between the two, the greater the problem. Problem solving in a total quality setting is not just putting out fires as they occur. Rather, it is one more way to make continual improvements in the workplace and its products or services to prevent recurrence of problems. This section contains a model for solving problems in ways that simultaneously lead to workplace improvements: the PDCA cycle.

The Plan-Do-Check-Adjust (PDCA) Cycle

This continual improvement model goes by several names. In Japan it is called the *Deming Cycle* after Dr. W. Edwards Deming, who introduced it to them. (Refer to Figure 10–1.) Deming himself referred to it as the *Shewhart Cycle* after its originator, Dr. Walter Shewhart. In the West it is commonly called the *PDCA cycle,* standing for plan-do-check-act. In this book we have taken the liberty to suggest that the letter *A* more correctly means *adjust.* Whatever we call it, the PDCA cycle consists of four major components, each of which can be subdivided into step-by-step activities. Deming disciple William W. Scherkenbach explains the model as follows:[1]

1. *Plan: develop a plan to improve.*   Even before problems occur, create a plan for improving your area of responsibility, particularly the processes in that area. Then, when problems occur, they can be handled within the context of Deming's model for continuous improvement. Developing such a plan involves completing the following steps:
    - Identify opportunities for improvement.
    - Document the current process.

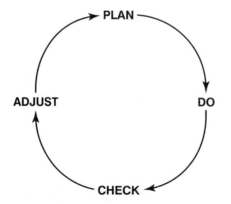

1. **Plan** action to improve process or product.
2. **Do** (implement) the plan.
3. **Check** to verify that planned results are achieved.
4. **Adjust**—determine changes necessary to achieve desired results.
5. **Repeat** the cycle.

Figure 10–1
The Deming (or PDCA) Cycle

- Create a vision of the improved process.
- Define the scope of the improvement effort.

2. *Do: carry out the plan.*   Implement the plan for improvement. The recommended approach is to first implement on a small scale over a specified period of time. This is the equivalent of developing and testing a prototype of a design before moving to full production.

3. *Study [Check]: examine the results.*   Examine and record the results achieved by implementing the plan. The recorded results form the basis for carrying out the steps in the next component.

4. *Act: adjust as necessary.*   Make adjustments as necessary based on what was learned in the previous component. Then repeat the cycle for the next planned improvement by returning to the first component of the model.

The PDCA cycle has evolved from that which Deming presented to his Japanese audience in the summer of 1950. The cycle started with *design the product*, followed in order by *production, sales,* and *market research.* His emphasis was on developing products that would be accepted in the world's markets, and that was precisely the need of the moment in Japan. In the design phase, he stressed finding out what is needed by potential customers, designing a product to meet the need, and planning sufficient production to validate the product's viability. That has become the Plan part of the cycle. The production plan was to be executed in the second quadrant of the cycle. That has become the Do phase. After producing the product, they were to sell it. Whether it sold well or poorly provided information on whether they had correctly chosen a product type. This has become the Check phase. Having sold the product, they were admonished to find out from their customers whether the product met their expectations and how it could be changed to better serve the customer. That has become the Act or Adjust phase. The concept was that a second cycle would commence immediately, taking into consideration everything that was learned from the first cycle; then a third, fourth, and so on, continuously applying information learned to redesign the product and find ways to make production more efficient, always with customer input as a very important input to the process.

Ishikawa's version is essentially that which we call the PDCA cycle. He specifies six steps:[2]

1. Determine goals and targets.
2. Determine methods of reaching goals. (Steps 1 and 2 constitute the Plan.)
3. Engage in education and training.
4. Implement work. (Steps 3 and 4 constitute the Doing.)
5. Check the effects of implementation (Check phase).
6. Take appropriate action (Action phase).

Where Deming's initial emphasis was on the product, Ishikawa's version seems to lend itself more clearly to processes as well as products. We know that Deming came to the same place, perhaps having been there all along, and there is no significant philosophical difference, only in the choice of words.

Application of PDCA in Problem Solving

Whether the issue is a single problem, or multiple simultaneous problems, the problem solving strategy outlined in Figure 10–2 applies.

1. Define the problem(s).

This means that the problem itself, not simply the symptoms of the problem, must be identified. Often the problem is not what it appears to be at first glance. Be as specific as possible.

1. **To Define the Problem:**
   - Establish a problem solving team.
   - Collect relevant facts.
   - Describe the problem.

2. **To Determine the Most Probable Cause(s):**
   - List all possible causes.
   - Refine the list.
   - Isolate the most probable cause(s).

3. **To Determine the Root Cause:**
   - Determine why the most probable cause occured.
   - Continue asking "why" until a root cause is determined.

4. **To Eliminate the Root Cause:**
   - Plan—develop a plan to eliminate the root cause.
   - Do—implement the plan.
   - Check—check to ensure that desired results are achieved.
   - Adjust—if desired results are not achieved, adjust the plan accordingly.
   - Repeat the PDCA cycle

Figure 10–2
Problem Solving Strategy for Total Quality

Establish a problem solving team.   The reason for using a team in solving problems is the same as that for using a team in any undertaking: no individual knows as much as a team. Team members have their own individual experiences, abilities, and perspectives. Consequently, the collective efforts of a team are typically more effective than the individual efforts of one person. Problem solving is best done by a team of people selected for their experience within the process involved, people who provide input to and take output from the process, and any others required for the specific skills and training necessary to find a solution.

Collect relevant facts.   There may be a temptation to jump right in and start solving the problem before it is identified. This is the "ready-fire-aim" approach. The better practice is to collect all available information about the problem before pursuing solutions. Two kinds of information can be collected: objective and subjective. *Objective information* is factual. *Subjective information* is open to interpretation. Nothing is wrong with collecting subjective information, as long as the following rules of thumb are adhered to. In fact, these rules apply to both subjective and objective information.

- Collect only information that pertains to the problem in question.
- Be thorough (it's better to have too much information than too little).
- Don't waste time recollecting information that is already on file.
- Allow sufficient time for thorough information collection, but set a definite time limit.
- Use the Quality Tools explained in Chapter 9.

To properly identify the problem, it must be broken down into its component parts as follows:

- Who is involved or affected?
- Where does it occur (location, process)?
- When did, or does, it occur?

- What happens when it occurs? What is the severity of impact?
- How often does it occur?

With this information in hand the team has what it needs to separate the real problem from the problem's symptoms.

Describe the problem. The team should define the problem in clear, concise terms, for example: machining process failed to apply proper finish to three lots of part number 6704-3.

2. Determine the most probable cause(s).

List possible causes. Identifying possible causes is a critical step in the process. To develop the list of all possible causes of the problem, the team should use a technique such as brainstorming. The brainstorming session should list every idea presented, not just the ones that sound most logical. The list should be all inclusive. Figure 10–3 is such a list.

Refine the list of possible causes. Once the team has listed every possible cause it can think of, the list may be refined by eliminating those that do not fit the circumstances or are illogical, or are found to be impossible.

Isolate the most probable cause. An effective tool for isolating the most probable causes of a particular problem is the cause-and-effect diagram introduced in Chapter 9. A cause-and-effect diagram for our hypothetical problem is illustrated in Figure 10–4. You will notice that the diagram resembles the bones of a fish, hence the name, "fishbone diagram," which is often used for it. The six bones on this particular diagram represent the customary six major groupings of causes: Machine (equipment), Manpower (operator), Materials, Measurement, Methods, and Environment. The head of the fish represents the problem. All causes of workplace problems fall into one of these major groupings. The team assigns each possible cause from its list to the appropriate bone (category). Through group discussion and investigation the team tests each possible cause against the problem definition. In doing this there are three possibilities: the cause will fully explain the problem, the cause will partially explain the problem, or the cause will not explain the problem. A cause that can fully explain the problem is a likely candidate to be the most probable cause. If more than one cause fully explains the problem, there may be more than one cause to the problem.

3. Determine the root cause.

Find out what is behind the most probable cause (i.e., Why did it happen?) Even with the most probable cause of the problem identified, the team must look further to be

---

List of Possible Causes for Failure to Apply Proper Finish to PN 6704-3

Measurement accuracy	Material suitability	Noise
Operator vision	Machine capability	Machine adjustment
Availability of procedure	Measurement	Operator skills
Lighting	repeatability	Machine maintenance
Material hardness	Operator availability	Procedure
Instrument capability	Instrument calibration	understandability
Machine repair	Difficulty of procedure	Operator training
Use of correct tools(s)	Temperature	Procedure thoroughness

Figure 10–3
List of Possible Causes From Brainstorming

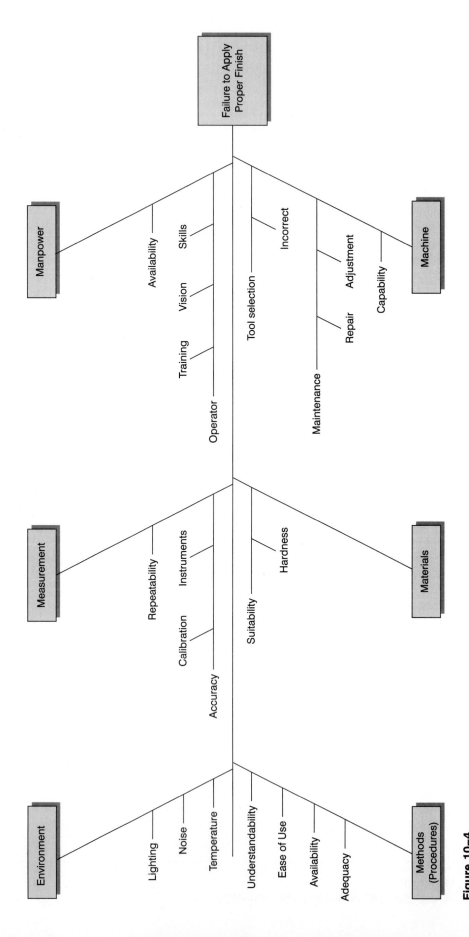

Figure 10–4

Cause-and-Effect Diagram: Failure to Apply Proper Finish

certain it has isolated the root cause. For example, the most probable cause may have been "operator error." However, operator error is not a root cause, and no amount of operator counseling will guarantee that the problem will not recur. We must determine why the operator made the error. The root cause might be a procedure that is difficult to follow, the failure of management to provide the necessary operator training, or any of a number of other factors. If the cause of the error was a procedure that was difficult to understand or follow, then we should ask, why was the procedure difficult to follow? It may simply have been a case of poor writing. For example, if it turns out that this particular procedure alone among all the organization's procedures was poorly written, then the root cause was an inadequate procedure. However, if the procedure was difficult because the organization's procedures were written at a level above the typical operator's education, then the organization's policy for its procedures that specifies a 12th-grade reading level may be the root cause.

If this should prove to be the case, all the other procedures should be considered problems-about-to-happen, and should also be reviewed to be certain that the users do not have difficulty understanding or following them. Doing this can prevent problems before they happen.

What we did here to arrive at a root cause was to repeatedly ask "Why?".

■ Why did the finishing process fail to apply the proper finish?

> Causes could have included machines, methods, materials, measurement, manpower (operator), and environment. In this case, "operator error" was determined to be the most probable cause.

■ Why did the operator make the error?

> Causes could have included the same list as above. In this case the cause was a procedure that was difficult for the operator to understand or follow.

■ Why was the procedure difficult to understand or follow?

> Once again, the causes could have included factors from the same list. Being written in such a way as to be difficult for the operators was the case here.

■ Why was the procedure poorly written?

> The probable cause may be as simple as inadequate writing skills on the part of the person who wrote the procedure. Or it might have been the result of a company policy requiring that its procedures be written to a 12th-grade reading level, thus putting the procedures beyond the comprehension of many operators.

4. Apply the plan-do-check-adjust cycle.

Plan A plan of action must be developed to eliminate the root cause of the problem, thus ensuring the same problem never happens again. If the root cause of the problem was, in our case, simply a writing skills problem involving only the procedure in question, then rewriting the procedure will eliminate the root cause. However, if all the organization's procedures were written over the heads of many employees, the company policy is the root cause. If this is true, then in addition to rewriting the procedure in question, the plan must do several things:

■ Require the offending procedure to be rewritten
■ Require a change to the company policy
■ Require that all other procedures be tested for understandability and ease of following
■ Require rewriting of any procedures that do not measure up for ease of understanding and following by the procedures' users

The plan must specify what is to be done, and by whom, to eliminate the root cause, as we have just discussed.

Do   Implement the Plan. Let's say that only this particular procedure was poorly written. Therefore the plan calls for the procedure to be rewritten so that it is easier to read, comprehend, and follow. This is the step in which the new procedure is written and put into service.

Check   In this step the team must determine whether the implemented plan produces the desired results. The team must obtain feedback from the procedure's users by observation or by inquiry. The operators may confirm that they no longer have difficulty understanding or following the procedure, or they may not yet be satisfied. When the outcome is not quite what was desired, that information is used in the next phase of the PDCA cycle.

Adjust   If the desired results have not been achieved, the team must determine what adjustments need to be made in the plan for the next PDCA cycle.

Repeat   The PDCA cycle is repeated, revising the plan to accommodate the necessary adjustments. Do (implement) the revised plan. Check the results of the revised plan. Identify any further adjustments that should be made, and go through another cycle.

PROBLEM-SOLVING AND DECISION-MAKING TOOLS

In arriving at the possible answers to the "why" questions, brainstorming and cause-and-effect diagrams may be effectively employed. Use of the other quality tools, Pareto, scatter diagrams, histograms, and run charts is also recommended as appropriate for the data. Stratification is a technique that can lead to root causes. Flowcharts are also invaluable in many situations involving processes. For detailed information about each of these tools, refer to Chapter 9.

DECISION MAKING FOR TOTAL QUALITY

One of the world's leading experts on business leadership, Dr. John P. Kotter, Konosuke Matsushita Professor of Leadership at the Harvard Business School states:[3]

1. Today's business environment demands more large-scale change via new strategies, reengineering, restructuring, mergers, acquisitions, downsizing, new product or market development, etc.
2. Decisions made inside the firm
   - are based on bigger, more complex, more emotionally charged issues.
   - are made more quickly.
   - are made in a less certain environment.
   - require more sacrifice from those implementing the decisions.
3. A new decision-making process
   - is required because no one individual has the information needed to make all major decisions, or the time and credibility needed to convince lots of people to implement the decisions.
   - must be guided by a powerful coalition that can act as a team.

Dr. Kotter lends credence to the thesis that organizational decisions can no longer be made the way we have been making them for the last 100 years. As he points out, today's business decisions cannot be made without sufficient knowledge of all the relevant factors, which often means that the collective knowledge of the organization must be tapped. At the least, we must be smart in our decision making, or we may find ourselves on the path to ruin.

All people make decisions. Some are minor. (What should I wear to work today? What should I have for breakfast?) Some are major. (Should I accept a job offer in another city? Should I buy a new house?) Regardless of the nature of the decision, decision making can be defined as follows:

> *Decision making* is the process of selecting one course of action from among two or more alternatives.

Decision making is a critical task in a total quality setting. Decisions play the same role in an organization that fuel and oil play in an automobile engine: they keep it running. The work of an organization cannot proceed until decisions are made.

Consider the following example. Because a machine is down, the production department at DataTech, Inc., has fallen behind schedule. With this machine down, DataTech cannot complete an important contract on time without scheduling at least 75 hours of overtime. The production manager faces a dilemma. On the one hand, no overtime was budgeted for the project. On the other hand, there is substantial pressure to complete this contract on time because future contracts with this client may depend on it. The manager must make a decision.

In this case, as in all such situations, it is important to make the right decision. But how do managers know when they have made the right decision? In most cases, there is no single right choice. If there were, decision making would be easy. Typically several alternatives exist, each with its own advantages and disadvantages.

For example, in the case of DataTech, Inc., the manager had two alternatives: authorize 75 hours of unbudgeted overtime or risk losing future contracts. If the manager authorizes the overtime, his or her company's profit for the project in question will suffer, but its relationship with a client may be protected. If the manager refuses to authorize the overtime, the company's profit on this project will be protected, but the relationship with this client may be damaged. These and other types of decisions must be made all the time in the modern workplace.

Managers should be prepared to have their decisions evaluated and even criticized after the fact. Although it may seem unfair to conduct a retrospective critique of decisions that were made during the heat of battle, having one's decisions evaluated is part of accountability, and it can be an effective way to improve a manager's decision-making skills.

Evaluating Decisions

There are two ways to evaluate decisions. The first is to examine the results. In every case when a decision must be made, there is a corresponding result. That result should advance an organization toward the accomplishment of its goals. To the extent that it does, the decision is usually considered a good decision. Managers have traditionally had their decisions evaluated based on results. However, this is not the only way that decisions should be evaluated. Regardless of results, it is wise also to evaluate the process used in making a decision. Positive results can cause a manager to overlook the fact that a faulty process was used, and, in the long run, a faulty process will lead to negative results more frequently than to positive.

For example, suppose a manager must choose from among five alternatives. Rather than collect as much information as possible about each, weigh the advantages and disadvantages of each, and solicit informed input, suppose the manager chooses randomly. He or she has one chance in five of choosing the best alternative. Such odds occasionally produce a positive result, but typically they don't. This is why it is important to examine the process as well as the result, not just when the result is negative but also when it is positive.

THE DECISION-MAKING PROCESS

Decision making is a process. For the purpose of this textbook, the decision-making process is defined as follows:

> The *decision-making process* is a logically sequenced series of activities through which decisions are made.

Numerous decision-making models exist. Although they appear to have major differences, all involve the various steps shown in Figure 10–5 and discussed next.

Identify or Anticipate the Problem

If managers can anticipate problems, they may be able to prevent them. Anticipating problems is like driving defensively: never assume anything. Look, listen, ask, and sense. For example, if you hear through the grapevine that a team member's child has been severely injured and hospitalized, you can anticipate the problems that may occur. She is likely to be absent, or if she does come to work, her pace may be slowed. The better managers know their employees, technological systems, products, and processes, the better able they will be to anticipate problems.

Gather the Facts

Even the most perceptive managers will be unable to anticipate all problems or to understand intuitively what is behind them. For example, suppose a manager notices a "who cares?" attitude among team members. This manager might identify the problem as poor morale and begin trying to improve it. However, he or she would do well to gather the facts first to be certain of what is behind the negative attitudes. The underlying cause(s) could come from a wide range of possibilities: an unpopular management policy, dissatisfaction with the team leader, a process that is ineffective, problems at home, and so forth. Using the methods and tools described earlier in this chapter and in Chapter 9, the manager should separate causes from symptoms and determine the root cause of the problem. Only by doing so will the problem be permanently resolved. The inclusion of this step makes possible *management by facts*—a cornerstone of the total quality philosophy.

It should be noted that the factors that might be at the heart of a problem include not only those for which a manager is responsible (policies, processes, tools, training,

Figure 10–5
Decision-Making Model

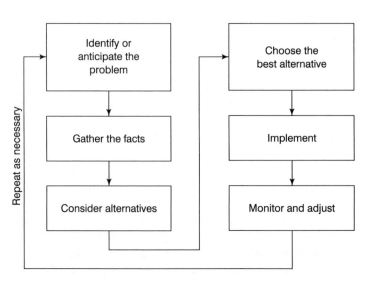

personnel assignment, etc.) but possibly also ones beyond the manager's control (personal matters, regulatory requirements, market and economic influences, etc.). For those falling within the manager's domain of authority, he or she must make sound, informed decisions based on fact. For the others, the organization has to adapt.

Consider Alternatives

This involves two steps: (a) list all of the various alternatives available, and (b) evaluate each alternative in light of the facts. The number of alternatives identified in the first step will be limited by several factors. Practical considerations, the manager's range of authority, and the cause of the problem will all limit a manager's list of alternatives. After the list has been developed, each entry is evaluated. The main criterion against which alternatives are evaluated is the desired outcome. Will the alternative being considered solve the problem? If so, at what cost?

Cost is another criterion used in evaluating alternatives. Alternatives always come with costs, which might be expressed in financial terms, in terms of employee morale, in terms of the organization's image, or in terms of a client's goodwill. Such costs should be considered when evaluating alternatives. In addition to applying objective criteria and factual data, managers will also need to apply their judgment and experience when considering alternatives.

Choose the Best Alternative, Implement, Monitor, and Adjust

After all alternatives have been considered, one must be selected and implemented, and after an alternative has been implemented, managers must monitor progress and adjust appropriately. Is the alternative having the desired effect? If not, what adjustments should be made? Selecting the best alternative is never a completely objective process. It requires study, logic, reason, experience, and even intuition. Occasionally, the alternative chosen for implementation will not produce the desired results. When this happens and adjustments are not sufficient, it is important for managers to cut their losses and move on to another alternative. (The PDCA cycle will help managers with this process).

Managers should avoid falling into the ownership trap. This happens when they invest so much ownership in a given alternative that they refuse to change even when it becomes clear the idea is not working. This can happen at any time but is more likely when a manager selects an alternative that runs counter to the advice he or she has received, is unconventional, or is unpopular. The manager's job is to solve the problem. Showing too much ownership in a given alternative can impede the ability to do so.

OBJECTIVE VERSUS SUBJECTIVE DECISION MAKING

All approaches to decision making fall into one of two categories: objective or subjective. Although the approach used by managers in a total quality setting may have characteristics of both, the goal is to minimize subjectivity and maximize objectivity. The approach most likely to result in a quality decision is the objective approach.

Objective Decision Making

The objective approach is logical and orderly. It proceeds in a step-by-step manner and assumes that managers have the time to systematically pursue all steps in the decision-making process (see Figure 10–6). It also assumes that complete and accurate information is available and that managers are free to select what they feel is the best alternative.

Measured against these assumptions, it can be difficult to be completely objective when making decisions. Managers don't always have the luxury of time and complete

Figure 10–6
Factors That Contribute to Objective Decision Making

• Freedom to select the best alternative
• Complete accurate information (Facts)
• Time

information. This does not mean that objectivity in decision making should be considered impossible. Managers should be as objective as possible. However, it is important to understand that the day-to-day realities of the workplace may limit the amount of time and information available. When this is the case, objectivity can be affected.

Subjective Decision Making

Whereas objective decision making is based on logic and complete, accurate information, subjective decision making is based on intuition, experience, and incomplete information. This approach assumes decision makers will be under pressure, short on time, and operating with only limited information. The goal of subjective decision making is to make the best decision possible under the circumstances. In using this approach, the danger always exists that managers might make quick, knee-jerk decisions based on no information, misinterpretation of available information, or no input from other sources. The subjective approach does not give managers license to make sloppy decisions. If time is short, the little time available should be used to list and evaluate alternatives. If information is incomplete, use as much information as is available. Subjective decision making is an anathema in the total quality context, and it should be avoided whenever possible.

SCIENTIFIC DECISION MAKING AND PROBLEM SOLVING

As explained in the previous section, sometimes decisions must be made subjectively. However, through good management and leadership, such instances should and can be held to a minimum. One of the keys to success in a total quality setting is using a scientific approach in making decisions and solving problems.[4] Peter R. Scholtes explains Joseph M. Juran's 85/15 rule as follows:

> There is a widely held belief that an organization would have few, if any, problems if only workers would do their jobs correctly. As Dr. Joseph M. Juran pointed out years ago, this belief is incorrect. In fact, the potential to eliminate mistakes and errors lies mostly in improving the "systems" through which work is done, not in changing the workers. This observation has evolved into the rule of thumb that at least 85% of problems can only be corrected by changing systems (which are largely determined by management) and less than 15% are under a worker's control—and the split may lean even more towards the system.[5]

Decision makers in a total quality setting should understand this rule. It is one of the fundamental premises underlying the need for scientific decision making.

Scholtes provides the following rationale for scientific decision making:

> The core of quality improvement methods is summed up in two words: scientific approach. Though this may sound complicated, a scientific approach is really just a systematic way for individuals and teams to learn about processes. It means agreeing to make decisions based on data rather than hunches, to look for root causes of problems rather than react to superficial symptoms, to seek permanent solutions rather than rely on quick fixes. A scientific approach can, but does not always, involve using sophisticated statistics, formulas, and experiments. These tools enable us to go beyond band-aid methods that merely cover up problems to find permanent, upstream improvements.[6]

Complexity and the Scientific Approach

In the language of scientific decision making, complexity means nonproductive, unnecessary work that results when organizations try to improve their processes without first developing a systematic plan.[7] Several different types of complexity exist, including the following: errors and defects, breakdowns and delays, inefficiencies, and variation. Decision makers should keep the Pareto Principle in mind when attempting to apply the scientific approach:

> This principle is sometimes called the 80/20 rule: 80% of the trouble comes from 20% of the problems. Though named for turn-of-the-century economist Vilfredo Pareto, it was Dr. Juran who applied the idea to management. Dr. Juran advises us to concentrate on the 'vital few' sources of problems and not be distracted by those of lesser importance.[8]

Errors and Defects

Errors cause defects and defects reduce competitiveness. When a defect occurs, one of two things must happen: the part or product must be scrapped altogether, or extra work must be done to correct the defect. Waste or extra work that results from errors and defects adds cost to the product without adding value.

Breakdowns and Delays

Equipment breakdowns delay work, causing production personnel either to work overtime or to work faster to catch up. Overtime adds cost to the product without adding value. When this happens, the organization's competitors gain an unearned competitive advantage. When attempts are made to run a process faster than its optimum rate, an increase in errors is inevitable.

Inefficiency

Inefficiency means using more resources (time, material, movement, or something else) than necessary to accomplish a task. Inefficiency often occurs because organizations fall into the habit of doing things the way they have always been done without ever asking why.

Variation

In a total quality setting, consistency and predictability are important. When a process runs consistently, efforts can begin to improve it by reducing process variations, of which there are two kinds:

- *Common-cause variation* is the result of the sum of numerous small sources of natural variation that are always part of the process.
- *Special-cause variation* is the result of factors that are not part of the process and that occur only in special circumstances, such as a shipment of faulty raw material or the involvement of a new, untrained operator.

The performance of a process that operates consistently can be recorded and plotted on a control chart such as the one in Figure 10–7. The sources of the variation in this figure that fall within the control limits are likely to be common causes. The sources of variation in this figure that fall outside the control limits are likely to be special-causes. In making decisions about the process in question, it is important to separate common and special causes of variation.

Commenting on variation, Scholtes says:

> If you react to common-cause variation as if it were due to special causes, you will only make matters worse and increase variation, defects, and mistakes. If you fail to notice the appearance of a special cause, you will miss an opportunity to search out and eliminate a source of problems.[9]

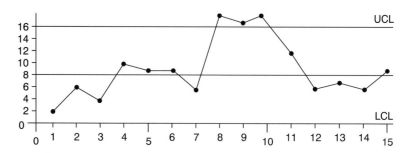

Figure 10–7
Control Chart

The concept of using control charts and statistical data in decision making is discussed in greater depth in Chapters 9 and 11.

EMPLOYEE INVOLVEMENT IN DECISION MAKING

Chapter 5 showed how employee involvement and empowerment can improve decision making. Employees are more likely to show ownership in a decision they had a part in making. Correspondingly, they are more likely to support a decision for which they feel ownership. There are many advantages to be gained from involving employees in decision making, as was shown in Chapter 5. There are also factors that, if not understood and properly handled, can lead to problems.

Advantages of Employee Involvement

Involving employees in decision making can have a number of advantages. It can result in a more accurate picture of what the problem really is and a more comprehensive list of potential solutions. It can help managers do a better job of evaluating alternatives and selecting the best one to implement.

Perhaps the most important advantages are gained after the decision is made. Employees who participate in the decision-making process are more likely to understand and accept the decision and have a personal stake in making sure the alternative selected succeeds.

Potential Problems with Employee Involvement

Involving employees in decision making can lead to problems. The major potential problem is that it takes time, and managers do not always have time. Other potential problems are that it takes employees away from their jobs and that it can result in conflict among team members. Next to time, the most significant potential problem is that employee involvement can lead to democratic compromises that do not necessarily represent the best decision. In addition, disharmony can result when a decision maker rejects the advice of the group.

Nevertheless, if care is taken, managers can gain all of the advantages while avoiding the potential disadvantages associated with employee involvement in decision making. Several techniques are available to help increase the effectiveness of group involvement. Prominent among these are brainstorming, the nominal group technique (NGT), and quality circles. Be particularly wary of the dangers of *groupthink* and *groupshift* in group decision making as outlined in Chapter 5.

ROLE OF INFORMATION IN DECISION MAKING

Information is a critical element in decision making. Although having accurate, up-to-date, comprehensive information does not guarantee a good decision, lacking such information can guarantee a bad one. The old saying that knowledge is power applies in decision making—particularly in a competitive situation. To make decisions that will help their organizations be competitive, managers need timely, accurate information.

Information can be defined as data that have been converted into a useable format that is relevant to the decision-making process.

Data that are relevant to decision making are those that might have an impact on the decision. Communication is a process that requires a sender, a medium, and a receiver. In this process, information is what is provided by the sender, transmitted by the medium, and received by the receiver. For the purpose of this chapter, decision makers are receivers of information who base decisions at least in part on what they receive.

Advances in technology have ensured that the modern manager can have instant access to information. Computers and telecommunications technology give decision makers a mechanism for collecting, storing, processing, and communicating information quickly and easily. The quality of the information depends on people (or machines) receiving accurate data, entering it into technological systems, and updating it continually. This dependence on accurate information gave rise to the expression "garbage in/garbage out" that is now associated with computer-based information systems. The saying means that information provided by a computer-based system can be no better than the data put into the system.

Data versus Information

Data for one person may be information for another. The difference is in the needs of the individual. Managers' needs are dictated by the types of decisions they make. For example, a computer printout listing speed and feed rates for a company's machine tools would contain valuable information for the production manager; the same printout would be just data to the warehouse manager. In deciding on the type of information they need, decision makers should ask themselves these questions:

- What are my responsibilities?
- What are my organizational goals?
- What types of decisions do I have to make relative to these responsibilities and goals?

Value of Information

Information is a useful commodity. As such it has value. Its value is determined by the needs of the people who will use it and the extent to which the information will help them meet their needs. Information also has a cost. Because it must be collected, stored, processed, continually updated, and presented in a usable format when needed, information can be expensive. This fact requires managers to weigh the value of information against its cost when deciding what information they need to make decisions. It makes no sense to spend $100 on information to help make a $10 decision.

Amount of Information

An old saying holds that a manager can't have too much information. This is no longer true. With advances in information technologies, not only can managers have too much information, but they frequently do. This phenomenon has come to be known as *information overload*, the condition that exists when people receive more information

Figure 10–8
Information Overload

> **Problems caused by information overload include:**
>
> - Too much attention given to unimportant matters
> - Too little attention given to important details
> - Confusion
> - Frustration
> - Unnecessary, unproductive delays

than they can process in a timely manner. The phrase "in a timely manner" means in time to be useful in decision making (see Figure 10–8).

To avoid information overload, apply a few simple strategies. First, examine all regular reports received. Are they really necessary? Do you receive daily or weekly reports that would meet your needs just as well if provided on a monthly basis? Do you receive regular reports that would meet your needs better as exception reports? In other words, would you rather receive reports every day that say everything is all right or occasional reports when there is a problem? The latter approach is reporting by exception and can cut down significantly on the amount of information that managers must absorb.

Another strategy for avoiding information overload is formatting for efficiency. This involves working with personnel who provide information. If your organization has a management information systems (MIS) department, ensure that reports are formatted for your convenience rather than theirs. Decision makers should not have to wade through reams of computer printouts to locate the information they need. Nor should they have to become bleary-eyed reading rows and columns of tiny figures. Work with MIS personnel to develop an efficient report form that meets your needs. Also, have that information presented graphically whenever possible.

Finally, make use of on-line, on-demand information retrieval. In the modern workplace, most reports are computer generated. Rather than relying on periodic printed reports, learn to retrieve information from the MIS database when you need it (on demand) using a computer terminal or a networked personal computer (on-line).

USING MANAGEMENT INFORMATION SYSTEMS (MIS)

The previous section contained references to management information systems and MIS personnel.

> A *management information system* is a system used to collect, store, process, and present information used by managers in decision making.

In the modern workplace, a management information system is typically a computer-based system. A management information system has three major components: hardware, software, and people. *Hardware* consists of the computer—be it a mainframe, mini-, or microcomputer—all of the peripheral devices for interaction with the computer, and output devices such as printers and plotters.

Software is the component that allows the computer to perform specific operations and process data. It consists primarily of computer programs but also includes the database, files, and manuals that explain operating procedures. *Systems software* controls the basic operation of the system. *Applications software* controls the processing of data for specific computer applications (word processing, CAD/CAM, computer-assisted process planning, spreadsheets, etc.).

A *database* is a broad collection of data from which specific information can be drawn. For example, a company might have a personnel database in which many different items of information about its employees are stored. From this database can be drawn a variety of different reports—such as printouts of all employees in order of employment date, by job classification, or by ZIP code. Data are kept on computer disks or tape on which they are stored under specific groupings or file names.

The most important MIS component is the people component. It consists of the people who manage, operate, maintain, and use the system. Managers who depend on a management information system for part of the information needed to make decisions are users.

Managers should not view a management information system as the final word in information. Such systems can do an outstanding job of providing information about predictable matters that are routine in nature. However, many of the decisions managers have to make concern problems that are not predictable and for which data are not tracked. For this reason, it is important to have sources other than the management information system from which to draw information.

CREATIVITY IN DECISION MAKING

The increasing pressures of a competitive marketplace are making it more and more important for organizations to be flexible, innovative, and creative in decision making. To survive in an unsure, rapidly changing marketplace, organizations must be able to adjust rapidly and change directions quickly. To do so requires creativity at all levels of the organization.

Creativity Defined

Like leadership, creativity has many definitions, and viewpoints vary about whether creative people are born or made. For the purposes of modern organizations, creativity can be viewed as an approach to problem solving and decision making that is imaginative, original, and innovative. Developing such perspectives requires that decision makers have knowledge and experience regarding the issue in question.

Creative Process

According to H. Von Oech, the creative process proceeds in four stages: preparation, incubation, insight, and verification.[10] What takes place in each of these stages is summarized as follows:

- *Preparation* involves learning, gaining experience, and collecting/storing information in a given area. Creative decision making requires that the people involved be prepared.
- *Incubation* involves giving ideas time to develop, change, grow, and solidify. Ideas incubate while decision makers get away from the issue in question and give the mind time to sort things out. Incubation is often a function of the subconscious mind.
- *Insight* follows incubation. It is the point in time when a potential solution falls in place and becomes clear to decision makers. This point is sometimes seen as a moment of inspiration. However, inspiration rarely occurs without having been preceded by perspiration, preparation, and incubation.
- *Verification* involves reviewing the decision to determine whether it will actually work. At this point, traditional processes such as feasibility studies and cost–benefit analyses are used.

Factors That Inhibit Creativity

A number of factors can inhibit creativity. Some of the more prominent of these are as follows:[11]

- *Looking for just one right answer.*   Seldom is there just one right solution to a problem.
- *Focusing too intently on being logical.*   Creative solutions sometimes defy logic and conventional wisdom.
- *Avoiding ambiguity.*   Ambiguity is a normal part of the creative process. This is why the incubation step is so important.
- *Avoiding risk.*   When organizations don't seem to be able to find a solution to a problem, it often means decision makers are not willing to give an idea a chance.
- *Forgetting how to play.*   Adults sometimes become so serious they forget how to play. Playful activity can stimulate creative ideas.
- *Fear of rejection or looking foolish.*   Nobody likes to look foolish or feel rejection. This fear can cause people to hold back what might be creative solutions.
- *Saying "I'm not creative."*   People who decide they are not creative won't be. Any person can think creatively and can learn to be even more creative.

Helping People Think Creatively

In the age of high technology and global competition, creativity in decision making and problem solving is critical. Although it is true that some people are naturally more creative than others, it is also true that any person can learn to think creatively. In the modern workplace, the more people who think creatively, the better. Darrell W. Ray and Barbara L. Wiley recommend the following strategies for helping employees think creatively:[12]

- *Idea vending.*   This is a facilitation strategy. It involves reviewing literature in the field in question and compiling files of ideas contained in the literature. Periodically, circulate these ideas among employees as a way to get people thinking. This will facilitate the development of new ideas by the employees. Such an approach is sometimes called stirring the pot.
- *Listening.*   One of the factors that causes good ideas to fall by the wayside is poor listening. Managers who are perpetually too hurried to listen to employees' ideas do not promote creative thinking. On the contrary, such managers stifle creativity. In addition to listening to the ideas, good and bad, of employees, managers should listen to the problems employees discuss in the workplace. Each problem is grist for the creativity mill.
- *Idea attribution.*   A manager can promote creative thinking by subtly feeding pieces of ideas to employees and encouraging them to develop the idea fully. When an employee develops a creative idea, he or she gets full attribution and recognition for the idea. Time may be required before this strategy pays off, but with patience and persistence it can help employees become creative thinkers.

How does a football team that is no better than its opponent beat that opponent? Often, the key is more creative game planning, play calling, and defense. This phenomenon also occurs in the workplace every day. The organization that wins the competition in the marketplace is often the one that is the most creative in decision making and problem solving.

ENDNOTES

1. William W. Scherkenbach, *Deming's Road to Continual Improvement* (Knoxville, TN: SPC, 1991), 63–66.
2. Kaoru Ishikawa, *What Is Total Quality Control? The Japanese Way* (Upper Saddle River, NJ: Prentice Hall, 1987), 59.
3. John P. Kotter, *Leading Change* (Boston: Harvard Business School Press, 1996), 56.
4. Joseph M. Juran, *Juran on Leadership for Quality: An Executive Handbook* (New York: Free Press, 1989), 163.
5. Peter R. Scholtes, *The Team Handbook* (Madison, WI: Joiner Associates, 1992), section 2-8.
6. Ibid.
7. This section based on Scholtes 2-9–2-15.
8. Scholtes, 2-9.
9. Scholtes, 2-13.
10. H. Von Oech, *A Whack on the Side of the Head* (New York: Warner, 1983), 77.
11. Von Oech, 77.
12. Darell W. Ray and Barbara L. Wiley, "How to Generate New Ideas," *Management for the 90s: A Special Report from Supervisory Management* (Saranac Lake, NY: American Management Association, 1991), 6–7.

CHAPTER ELEVEN

Statistical Process Control (SPC)

The origin of what is now called statistical process control (SPC) goes back to 1931 and Dr. Walter Shewhart's book *The Economic Control of Quality of Manufactured Product*. Shewhart, a Bell Laboratories statistician, was the first to recognize that industrial processes themselves could yield data, which, through the use of statistical methods, could signal that the process was in control or was being affected by special causes (causes beyond the natural, predictable variation). The control charts used today are based on Shewhart's work. These control charts are the very heart of SPC. What may not be as obvious is that Shewhart's work became the catalyst for the quality revolution in Japan[1] and the entire movement now called total quality. We tend to look at SPC as one piece of the whole total quality picture, and it is, but it is also the genesis of total quality.

Two very significant things have occurred in the SPC field over the last few years. First, many organizations have adopted SPC as a preferred way of controlling manufacturing processes. Much of this has come about as a result of the quality quest by first-tier companies, making it necessary to require that their second-tier suppliers practice SPC. We have seen this ripple down to at least the fourth tier. Nowhere is this more evident than in the auto industry. But even beyond the mandate by corporate customers, more and more small companies are using SPC as part of their quality and competitiveness initiatives.

The second big change we have seen is that SPC users have backed away from the shotgun approach, where every process, no matter how trivial or foolproof, had to have SPC charts. Four years ago we visited a North American semiconductor plant and were overwhelmed by the sheer numbers of SPC charts. Everywhere you looked you saw control charts. The plant proudly admitted to having over 900 processes under control charts. When we visited the same plant a couple of years later the picture was very different. You could still find control charts, but only where they offered real benefit. The company had discovered that about 800 of its original charts had not been worthwhile. Control charts were being used with those processes that needed them, and no more.

It is evident that this is the current thinking in industry. Don't waste time, energy and money with more control charts than you need. In those process applications where you do need them, the control chart is invaluable. For all the rest, it is just window dressing. The important thing is to know the difference.

STATISTICAL PROCESS CONTROL DEFINED

Although SPC is normally thought of in industrial applications, it can be applied to virtually any process. Everything done in the workplace is a process. All processes are affected by multiple factors. For example, in the workplace a process can be affected by the environment and the machines employed, the materials used, the methods (work instructions) provided, the measurements taken, and the manpower (people) who operate the process—the Five M's. If these are the only factors that can affect the process output, and if all of these are perfect—meaning the work environment facilitates quality work; there are no misadjustments in the machines; there are no flaws in the materials; there are totally accurate and precisely followed work instructions, accurate and repeatable measurements, and people who work with extreme care, following the work instructions perfectly, concentrating fully on their work—if all of these factors come into congruence, then the process will be in statistical control. This means that there are no special causes adversely affecting the process's output. Special causes are (for the time being, anyway) eliminated. Does that mean that 100% of the output will be perfect? No, it does not. Natural variation is inherent in any process, and it will affect the output. Natural variation is expected to account for roughly 3 out-of-limits parts in every 1,000 produced (the $\pm 3\sigma$ variation).

SPC does not eliminate all variation in the processes, but it does something that is absolutely essential if the process is to be consistent, and if the process is to be improved. SPC allows workers to separate the special causes of variation (e.g., environment and the Five M's) from the natural variation found in all processes. After the special causes have been identified and eliminated, leaving only natural variation, then the process is said to be in statistical control (or simply in control). When that state is achieved, the process is stable, and 99.73% of the output can be counted on to be within the statistical control limits. More important, improvement can begin. From this, we can develop a definition of statistical process control:

> Statistical process control (SPC) is a statistical method of separating variation resulting from special causes from variation resulting from natural causes, to eliminate the special causes and to establish and maintain consistency in the process, enabling process improvement.

RATIONALE FOR SPC

The rationale for SPC is much the same as that for total quality. It should not be surprising that the parallel exists, because it was Walter Shewhart's work that inspired the Japanese to invite W. Edwards Deming to help them get started in their quality program in 1949–1950. SPC was the seed from which the Japanese grew total quality.

The rationale for the Japanese to embrace SPC in 1950 was simple: a nation trying to recover from the loss of a costly war needed to export manufactured goods in order to import food for its people. The Asian markets once enjoyed by Japan had also been rendered extinct by the war. The remaining markets, principally North America, were unreceptive to Japanese products because of poor quality. If the only viable markets rejected Japanese products on the basis of quality, then Japanese manufacturers had to do something about their quality problem. This is why Shewhart's work interested them. This also is why they called on Deming, and later Joseph Juran, to help them. That the effort was successful is well documented and manifestly evident all over the world. Deming told the Japanese industrialists that if they would follow his teaching, they could become active players in the world's markets within 5 years. They actually made it in four.

The Western world may not be in the same crisis Japan experienced following World War II, but the imperative for SPC is no less crucial. When one thinks of quality products today, Japan still comes to mind first. Many of the finest consumer products in the world come from Japan. That includes everything from electronics and optical equipment to automobiles, although U. S. car manufacturers beginning with Ford have narrowed the quality gap. Fine automobiles are also produced in Europe; however, there is a serious price disparity. The European equivalent of a top-of-the-line Lexus or Infiniti made in Japan is the high-end Mercedes-Benz or BMW. These European luxury cars cost much more than their Japanese counterparts. In the United States, on the other hand, Cadillac and Lincoln are beginning to compete with Lexus and Infiniti on quality and price. Ford and General Motors are doing so by adopting such total quality strategies as SPC.

Most consumers are more interested in what Toyota, Honda, Ford, General Motors, Chrysler, and the others are doing at the middle and lower end of the automobile market. Until the early 1990s, the Japanese were the quality leaders in every strata of the automobile market. Cars made by Toyota, Nissan, Honda, Mazda, and Mitsubishi (including those produced in their North American factories) have been of consistently excellent quality. Now that manufacturers in the United States have adopted SPC and other total quality improvement strategies, the outcome of the race for quality leadership can no longer be predicted each new product year. Put simply, the rationale for Western manufacturers to embrace SPC has been not only to improve product quality and simultaneously reduce costs but also to improve their product image in order to compete successfully in the world's markets.

To understand how SPC can help accomplish this, it is necessary to examine five key points and understand how SPC comes into play in each one: control of variation, continuous improvement, predictability of processes, elimination of waste, and product inspection.

Rationale: Control of Variation

The output of a process that is operating properly can be graphed as a bell-shaped curve, as in Figure 11–1. The x-axis represents some measurement, such as weight or dimension, and the y-axis represents the frequency count of the measurements. The desired measurement value is at the center of the curve, and any variation from the desired value results in displacement to the left or right of the center of the bell. With no special causes acting on the process, 99.73% of the process output will be between the $\pm 3\sigma$ limits. (This is not a specification limit, which may be tighter or looser.) This de-

Figure 11–1
Frequency Distribution Curve:
Normal Curve

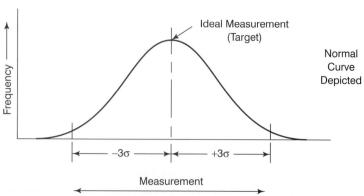

gree of variation about the center is the result of natural causes. The process will be consistent at this performance level as long as it is free of special causes of variation.

When a special cause is introduced, the curve will take a new shape, and variation can be expected to increase, lowering output quality. Figure 11–2 shows the result of a machine no longer capable of holding the required tolerance, or an improper work instruction. The bell is flatter, meaning that fewer parts produced by the process are at, or close to, the target, and more fall outside the original 3σ limits. The result is more scrap, higher cost, and inconsistency of product quality.

The curve of Figure 11–3 could be the result of input material from different vendors (or different batches) that is not at optimal specification. Again, a greater percentage of the process output will be displaced from the ideal, and more will be outside the original 3σ limits. The goal should be to eliminate the special causes to let the process operate in accordance with the curve shown in Figure 11–1, and then to improve the process, thereby narrowing the curve (see Figure 11–4).

When the curve is narrowed, more of the process output is in the ideal range, and less falls outside the original 3σ limits. Actually, each new curve will have its own 3σ limits. In the case of Figure 11–4, they will be much narrower than the original ones. If the original limits resulted in 2,700 pieces out of 1 million being scrapped, the improved process illustrated by Figure 11–4 might reduce that to 270 pieces, or even less, scrapped. Viewed from another perspective, the final product will be more consistently of high quality, and the chance for a defective product going to a customer is reduced by an order of magnitude.

Variation in any process is the enemy of quality. As we have already discovered, variation results from two kinds of causes: special causes and natural causes. Both kinds can be treated, but they must be separated so that the special causes—those associated with the Five M's and environment—can be identified and eliminated. After that is done, the processes can be improved, never eliminating the natural variation, but continuously narrowing its range, approaching perfection. It is important to understand why the special causes must first be eliminated. Until that happens the process will not be stable, and the output will include too much product that is unusable, therefore wasted. The process will not be dependable in terms of quantity or quality. In addition, it will be pointless to attempt improvement of the process, because one can never tell whether the improvement is successful—the results will be masked by the effect of any special causes that remain.

In this context, elimination of special causes is not considered to be process improvement, a point frequently lost on enthusiastic improvement teams. Elimination of

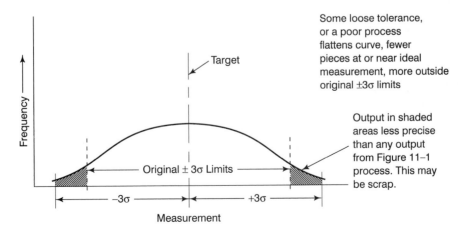

Figure 11–2
Frequency Distribution Curve: Process Not as Precise as Figure 11–1

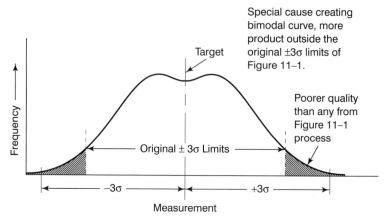

Figure 11–3
Bimodal Frequency Distribution Curve

special causes simply lets the process be whatever it will be in keeping with its natural variation. It may be good or bad or anything in between.

When thinking of SPC, most people think of control charts. If we wish to include the elimination of special causes as a part of SPC, as we should, then it is necessary to include more than the control chart in our set of SPC tools, because the control chart has limited value until the process is purged of special causes. If one takes a broad view of SPC, all of the statistical tools discussed in Chapter 9 should be included. Pareto charts, cause-and-effect diagrams, stratification, check sheets, histograms, scatter diagrams, and run charts are all SPC tools. Although the flowchart is not a statistical device, it is useful in SPC. The SPC uses of these techniques are highlighted later in this chapter. Suffice it to say for now that the flow diagram is used to understand the process better, the cause-and-effect diagram is used to examine special causes and how they impact the process, and the others are used to determine what special causes are at play and how important they are. The use of such tools and techniques makes possible the control of variation in any process to a degree unheard of before the introduction of SPC.

Rationale: Continuous Improvement

Continuous improvement is a key element of total quality. One talks about improvement of products, whatever they may be. In most cases, it would be more accurate to

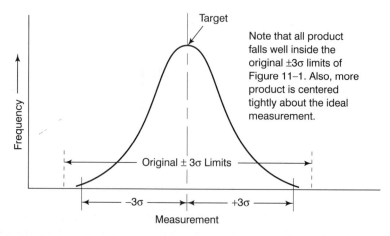

Figure 11–4
Frequency Distribution Curve: Narrowed (Less Variation) Relative to Figure 11–1

talk about continuous improvement in terms of processes than in terms of products and services. It is usually the improvement of processes that yields improved products and services. Those processes can reside in the engineering department where the design process may be improved by adding concurrent engineering and design-for-manufacture techniques, or in the public sector where customer satisfaction becomes a primary consideration. All people use processes, and all people are customers of processes. A process that cannot be improved is rare. We have not paid sufficient attention to our processes. Most people have only a general idea of what processes are, how they work, what external forces affect them, and how capable they are of doing what is expected of them. Indeed, outside the manufacturing industry, many people don't realize that their work is made up of processes.

Before a process can be improved, it is necessary to understand it, identify the external factors that may generate special causes of variation, and eliminate any special causes that are in play. Then, and only then, can we observe the process in operation and determine its natural variation. Once a process is in this state of statistical control, the process can be tracked, using control charts, for any trends or newly introduced special causes. Process improvements can be implemented and monitored. Without SPC, process improvement takes on a hit-or-miss methodology, the results of which are often obscured by variation stemming from undetected factors (special causes). SPC lets improvements be applied in a controlled environment, measuring results scientifically, and with assurance.

Rationale: Predictability of Processes

A customer asks whether a manufacturer can produce 300 widgets within a month. If so, the manufacturer will receive a contract to do so. The parts must meet a set of specifications supplied by the customer. The manufacturer examines the specifications and concludes that it can comply but without much margin for error. The manufacturer also notes that in a good month it has produced more than 300. So the order is accepted. Soon, however, the manufacturer begins having trouble with both the specifications and the production rate. By the end of the month only 200 acceptable parts have been produced. What happened? The same units with the same specification have been made before, and at a higher production rate. The problem is unpredictable processes. If the same customer had approached a firm that was versed in SPC, the results would have been different. The managers would have known with certainty their capability, and it would have been clear whether the customer's requirements could, or could not, be met. They would know because their processes are under control, repeatable, and predictable.

Few things in the world of manufacturing are worse than an undependable process. Manufacturing management spends half its time making commitments and the other half living up to them. If the commitments are made based on unpredictable processes, living up to them will be a problem. The only chance manufacturing managers have when their processes are not predictable is to be extra conservative when making commitments. Instead of keying on the best past performance, they look at the worst production month and base their commitments on that. This approach can relieve a lot of stress but can also lose a lot of business. In today's highly competitive marketplace (whether for a manufactured product or a service), organizations must have predictable, stable, consistent processes. This can be achieved and maintained through SPC.

Rationale: Elimination of Waste

Only in recent years have many manufacturers come to realize that production waste costs money. Scrap bins are still prominent in many factories. In the electronics industry, for example, it is not unusual to find that 25% of the total assembly labor cost in a product is expended correcting errors from preceding processes. This represents waste. Parts that are scrapped because they do not fit properly or are blemished represent waste. Parts that do not meet specifications are waste. To prevent defective products from going to customers, more is spent on inspection/reinspection. This, too, is waste. All of these situations are the result of some process not producing what was expected. In most cases, waste results from processes being out of control: processes are adversely influenced by special causes of variation. Occasionally, even processes that have no special causes acting on them are simply not capable of producing the expected result.

Two interesting things happen when waste is eliminated. The most obvious is that the cost of goods produced (or services rendered) is reduced—a distinct competitive advantage. At the same time, the quality of the product is enhanced.

Even when all parts are inspected, it is impossible to catch all the bad ones. When sampling is used, even more of the defective parts get through. When the final product contains defective parts, its quality has to be lower. By eliminating waste, a company reduces cost and increases quality. This suggests that Philip Crosby was too conservative: quality is not just free;[2] it pays dividends. This is the answer to the question of what happened to the Western industries that once led but then lost significant market share since the 1970s: total quality manufacturers simply built better products at competitive prices. These competitive prices are the result of the elimination of waste, not (as is often presumed) cheap labor. This was accomplished in Japan by applying techniques that were developed in the United States in the 1930s but ignored in the West after the crisis of World War II had passed. Specifically, through Japan's application of SPC and later the expansion of SPC to the broader concept of total quality, Japan went from a beaten nation to an economic superpower in just 30 years.

By concentrating on their production processes, eliminating the special causes as Shewhart and Deming taught, bringing the processes into statistical control, Japanese manufacturers could see what the processes were doing and what had to be done to improve them. Once in control, a relentless process improvement movement was started, one that is still ongoing a half century later—indeed, it is never finished. Tightening the bell curves brought ever-increasing product quality and ever-diminishing waste (nonconforming parts). For example, while U.S. automakers were convinced that to manufacture a more perfect transmission would be prohibitively expensive, the Japanese not only did it but also reduced its cost. In the early 1980s, the demand for a particular Ford transmission was such that Ford second-sourced a percentage of them from Mazda. Ford soon found that the transmissions manufactured by Mazda (to the same Ford drawings) were quieter, smoother, and more reliable than those produced in North America. Ford customers with Mazda transmissions were a lot happier than the others, as well. Ford examined the transmissions, and while they found that both versions were assembled properly with parts that met all specifications, the component parts of the Mazda units had significantly less variation piece to piece. Mazda employed SPC, and the domestic supplier did not. This demonstrated to Ford that the same design, when held to closer tolerances, resulted in a noticeably superior transmission that did not cost more. Shortly thereafter, Ford initiated an SPC program. To Ford's credit, their effort paid off. In 1993, the roles were reversed and Ford began producing transmissions for Mazda.

Statistical process control is the key to eliminating waste in production processes. It can do the same in virtually any kind of process. The inherent nature of process improvement is such that as waste is eliminated, the quality of the process output is correspondingly increased.

Rationale: Product Inspection

It is normal practice to inspect products as they are being manufactured (in-process inspection) and as finished goods (final inspection). Inspection requires the employment of highly skilled engineers and technicians, equipment that can be very expensive, factory space, and time. If it were possible to reduce the amount of inspection required, while maintaining or even improving the quality of products, money could be saved and competitiveness enhanced.

Inspection can be done on every piece (100% inspection) or on a sampling basis. The supposed advantage of 100% inspection is, of course, that any defective or non-conforming product will be detected before it gets into the hands of a customer (external or internal). The term *supposed advantage* is used because even with 100% inspection only 80% of the defects are found.[3] Part of the problem with 100% inspection is that human inspectors can become bored and, as a result, careless. Machine inspection systems do not suffer from boredom, but they are very expensive, and for many applications they are not a practical replacement for human eyes. It would be faster and less expensive if it were possible to achieve the same level of confidence by inspecting only 1 piece out of 10 (10% sampling) or 5 out of 100 (5% sampling) or even less.

Such sampling schemes are not only possible but accepted by such critical customers as the U.S. government (see the MIL-STD-105 and MIL-STD-414 series) and the automobile industry, but there is a condition: for sampling to be accepted, processes must be under control. Only then will the processes have the consistency and predictability necessary to support sampling. This is a powerful argument for SPC.

After supplier processes are under control and being tracked with control charts, manufacturers can back off the customary incoming inspection of materials, resorting instead to the far less costly procedure of periodically auditing the supplier's processes. SPC must first be in place, and the supplier's processes must be shown to be capable of meeting the customer's specifications.

This also applies internally. When a company's processes are determined to be capable of producing acceptable products, and after they are in control using SPC, the internal quality assurance organization can reduce its inspection and process surveillance efforts, relying to a greater degree on a planned program of process audit. This reduces quality assurance costs and, with it, the cost of quality.

CONTROL CHART DEVELOPMENT

Just as there must be many different processes, so must there be many types of control charts. The table in Figure 11–19 (presented later on page 189) lists the seven most commonly used control chart types. You will note that the first three are associated with measured values, or variables data. The other four are used with counted values, or attributes data. It is important as the first step in developing your control chart, to select the chart type that is appropriate for your data. The specific steps in developing control charts are different for variables data than for attributes data.

Control Chart Development for Variables Data (Measured Values)

Consider an example using $\bar{x}$ and R charts. $\bar{x}$ and R charts are individual, directly related graphs plotting the mean (average) of samples over time ($\bar{x}$), and the variation in each sample over time (R). The basic steps for developing a control chart for data with measured values are these:

1. Determine sampling procedure. Sample size may depend on the kind of product, the production rate, measurement expense, and likely ability to reveal changes in the process. Sample measurements are taken in subgroups of size (n), typically from 3 to 10. Sampling frequency should be often enough that changes in the process are not missed, but not so often as to mask slow drifts. If the object is to set up control charts for a new process, the number of subgroups for the initial calculations should be 25 or more. For existing processes that appear stable, that

number can be reduced to 10 or so, and sample size (n) can be smaller, say, 3 to 5.

2. Collect initial data of 100 or so individual data points in k subgroups of n measurements.
    -   Process must not be tinkered with during this time—let it run.
    -   Don't use old data—it may not be relevant to the current process.
    -   Take notes on anything that may have significance.
    -   Log data on a data sheet designed for control chart use.
3. Calculate the mean (average) values of data in each subgroup. ($\bar{x}$)
4. Calculate the data range for each subgroup. (R)
5. Calculate the average of the subgroup averages ($\bar{\bar{x}}$). This is the *process average* and will be the centerline for the $\bar{x}$-chart.
6. Calculate the average of the subgroup ranges ($\bar{R}$). This will be the centerline for the R-chart.
7. Calculate UCL and LCL (using a table of factors, such as the one shown in Figure 11–6). UCL and LCL represent $\pm 3\sigma$ limits *of the process averages* and are drawn as dashed lines on the control charts.
8. Draw the control chart to fit the calculated values.
9. Plot the data on the chart.

The reader should be careful to note that *upper control limit, lower control limit,* and *process average* are not arbitrary terms, nor are they the same as specifications and tolerances. They are statistically derived from the process's own running data. This cannot be emphasized too strongly. The problem is that if the control limits and process average are not statistically derived from the process, then it is impossible to know whether the process is in control and hence makes it enormously difficult to institute or validate process improvements. Yet we find many public and private organizations, and especially the military, using arbitrary or specification limits rather than statistical limits on their "control charts." Using this approach requires less work in setting up the charts, and they may look impressive to the uninitiated, but they are not control charts and can perform none of the functions of a control chart.

What follows is a step-by-step example of how to construct a control chart. First we have to collect sufficient data from which to make statistically valid calculations. This means we will usually have to take at least 100 data measurements in at least 10 subgroups, depending on the process, rate of flow, and so forth. The measurements should be made on samples close together in the process to minimize variation between the data points within the subgroups. However, the subgroups should be spread out in time to make visible the variation that exists between the subgroups.

The process for this example makes precision spacers that are nominally 100 millimeters thick. The process operates on a two-shift basis and appears to be quite stable. Fifty spacers per hour are produced. To develop a control chart for the process, we will measure the first 10 spacers produced after 9 A.M., 1 P.M., 5 P.M., and 9 P.M. We will do this for 3 days, for a total of 120 data points in 12 subgroups.

At the end of the 3 days, the data chart is as shown in Figure 11–5. The raw data are recorded in columns x_1 through x_{10}.

Next we calculate the mean (average) values for each subgroup. This is done by dividing the sum of x_1 through x_{10} by the number of data points in the subgroup.

$$\bar{x} = \Sigma x \div n$$

where n = the number of data points in the subgroup. The $\bar{x}$ values are found in the Mean Value column.

The average ($\bar{\bar{x}}$) of the subgroup averages ($\bar{x}$) is calculated by summing the values of $\bar{x}$ and dividing by the number of subgroups (k):

$$\bar{\bar{x}} = \Sigma \bar{x} \div k.$$

	Subgroup	Measured Values										Sum	Mean Value	Rng.
Date	#	x_1	x_2	x_3	x_4	x_5	x_6	x_7	x_8	x_9	x_{10}	Σx	$\overline{X}$	R
7/6	1	101	98	102	101	99	100	98	101	100	102	1002	100.2	4
7/6	2	103	100	101	98	100	104	102	99	101	98	1006	100.6	6
7/6	3	104	101	99	101	100	99	102	98	103	100	1007	100.7	5
7/6	4	96	99	102	99	101	102	98	100	99	97	993	99.3	6
7/7	5	99	102	100	99	103	101	102	98	100	100	1004	100.4	5
7/7	6	101	103	99	100	99	98	101	100	99	100	999	99.9	5
7/7	7	100	103	101	98	99	100	99	102	100	98	1000	100.0	5
7/7	8	97	101	102	100	99	96	99	100	103	98	995	99.5	7
7/8	9	102	97	100	101	103	98	100	102	99	101	1003	100.3	6
7/8	10	100	105	99	100	98	102	97	97	99	101	998	99.8	8
7/8	11	101	99	98	101	104	100	98	100	102	98	1001	100.1	6
7/8	12	100	103	101	98	99	100	100	99	98	102	1000	100.0	5
												Total 1200.8		68

$$k = 12, \quad \overline{\overline{X}} = 100.067, \quad \overline{R} = 5.667$$

Figure 11–5
Initial Data for Precision Spacer Process

In this case,

$$\overline{\overline{x}} = 1{,}200.8 \div 12$$
$$= 100.067$$

The range for each subgroup is calculated by subtracting the smallest value of x from the largest value of x in the subgroup.

$$R = (\text{maximum value of } x) - (\text{minimum value of } x)$$

Subgroup range values are found in the final column of Figure 11–5.
From the R values, calculate the average of the subgroup ranges.

$$\overline{R} = \Sigma R \div k$$

In this case,

$$\overline{R} = 68 \div 12$$
$$= 5.667$$

Next we calculate the UCL and LCL values for the $\overline{x}$ chart.

$$\text{UCL}_{\overline{x}} = \overline{\overline{x}} + A_2\overline{R} \qquad \text{LCL}_{\overline{x}} = \overline{\overline{x}} - A_2\overline{R}$$

At this point, you know the origin of all the values in these formulas except A_2. A_2 (as well as D_3 and D_4, used later) is from a factors table that has been developed for control charts (see Figure 11–6). The larger the value of A_2, the farther apart the upper and lower control limits $\text{UCL}_{\overline{x}}$ and $\text{LCL}_{\overline{x}}$ will be. A_2 may be considered to be a confidence factor for the data. The table shows that the value of A_2 decreases as the number of observations (data points) in the subgroup increases. It simply means that more data points make the calculations more reliable, so we don't have to spread the control limits so much. This works to a point, but the concept of diminishing returns sets in around $n = 15$.

Number of data points in subgroup	Factors for $\bar{x}$ charts	Factors for R charts	
		LCL	UCL
(n)	A_2	D_3	D_4
2	1.88	0	3.27
3	1.02	0	2.57
4	0.73	0	2.28
5	0.58	0	2.11
6	0.48	0	2.00
7	0.42	0.08	1.92
8	0.37	0.14	1.86
9	0.34	0.18	1.82
10	0.31	0.22	1.78
11	0.29	0.26	1.74
12	0.27	0.28	1.72
13	0.25	0.31	1.69
14	0.24	0.33	1.67
15	0.22	0.35	1.65
16	0.21	0.36	1.64
17	0.20	0.38	1.62
18	0.19	0.39	1.61
19	0.19	0.40	1.60
20	0.18	0.41	1.59

Figure 11–6
Factors Table for $\bar{x}$ and R Charts

Applying our numbers to the UCL and LCL formulas, we have this:

$$\text{UCL}_{\bar{x}} = 100.067 + (0.31 \times 5.667)$$
$$= 100.067 + 1.75677$$
$$= 101.82377$$
$$\text{LCL}_{\bar{x}} = 100.067 - 1.75677$$
$$= 98.31023$$

Now calculate the UCL and LCL values for the R chart.

$$\text{UCL}_R = D_4\bar{R} \qquad \text{LCL}_R = D_3\bar{R}$$

Like factor A_2 used in the $\bar{x}$ control limit calculation, factors D_3 and D_4 are found in Figure 11–6. Just as with A_2, these factors narrow the limits with subgroup size. With $n = 10$ in our example, $D_3 = 0.22$, and $D_4 = 1.78$. Applying the numbers to the LCL_R and UCL_R formulas, we have this:

$$\text{UCL}_R = 1.78 \times 5.667 \qquad \text{LCL}_R = 0.22 \times 5.667$$
$$= 10.08726 \qquad\qquad = 1.24674$$

At this point, we have everything we need to lay out the $\bar{x}$ and R charts (see Figure 11–7).

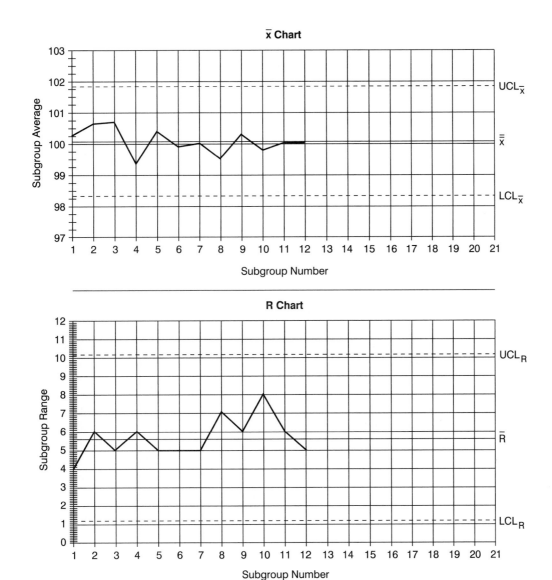

Figure 11–7
$\bar{x}$ and R Charts

The charts are laid out with y-axis scales set for maximum visibility consistent with the data that may come in the future. For new processes, it is usually wise to provide more y-axis room for variation and special causes. A rule of thumb is this:

■ (Largest individual value–smallest individual value) ÷ 2
■ Add that number to largest individual value to set the top of chart.
■ Subtract it from the smallest individual value to set the bottom of chart. (If this results in a negative number, set the bottom at zero.)

Upper and lower control limits are drawn on both charts as dashed lines, and $\bar{\bar{x}}$ and R centerlines are placed on the appropriate charts as solid lines. Then the data are plotted, subgroup average ($\bar{x}$) on the $\bar{x}$-chart, and subgroup range (R) on the R chart. We have arbitrarily established the time axis as 21 subgroups. It could be more or less, depending on the application. Our example requires space for 20 subgroups for a normal 5-day week.

Both charts in Figure 11–7 show the subgroup averages and ranges well within the control limits. The process seems to be in statistical control.

Figure 11–8
Chart for an Unstable Process

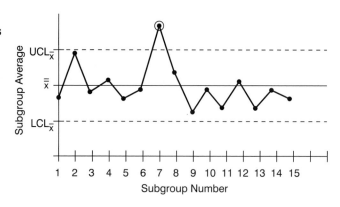

Suppose we had been setting up the charts for a new process (or one that was not as stable). We might have gotten a chart like the one in Figure 11–8.

Plotting the data shows that subgroup 7 was out of limits. This cannot be ignored, because the control limits have been calculated with data that included a nonrandom, special-cause event. We must determine and eliminate the cause. Suppose we were using an untrained operator that day. The operator has since been trained. Having established the special cause and eliminated it, we must purge the data of subgroup 7 and recalculate the process average ($\overline{\overline{x}}$) and the control limits. Upon recalculating, we may find that one or more of the remaining subgroup averages penetrates the new, narrower limits (as in Figure 11–9). If that happens, another iteration of the same calculation is needed to clear the data of any special-cause effects. We want to arrive at an initial set of charts that are based on valid data and in which the data points are all between the limits, indicating a process that is in statistical control (Figure 11–10). If after one or two iterations, all data points are not between the control limits, then we must stop. The process is too unstable for control chart application and must be cleared of special causes.

Control Chart Development for Attributes Data (Counted Data)

The *p*-chart

Attributes data are concerned not with measurement but with something that can be counted. For example, the number of defects is attributes data. Whereas the $\overline{x}$ and R charts are used for certain kinds of variables data, where measurement is involved, the *p*-chart is used for certain attributes data. Actually, the *p*-chart is used when the data are the *fraction defective of some set of process output*. It may also be shown as *per-*

Figure 11–9
The New, Narrower Limits Are Penetrated

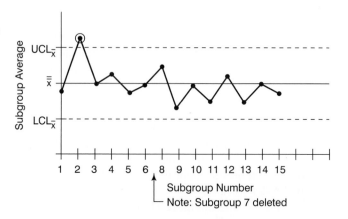

Figure 11–10
The Process Is in Statistical
Control

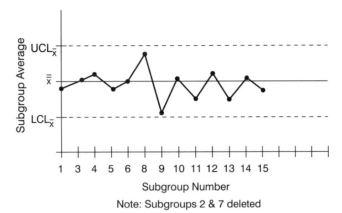

Note: Subgroups 2 & 7 deleted

centage defective. The points plotted on a *p*-chart are the fraction (or percentage) of defective pieces found in the sample of n pieces. The "Run Charts and Control Charts" section of Chapter 9 began with the example of a pen manufacturer. Now let's take that example to the next logical step and make a *p*-chart.

When we left the pen makers, they seemed to have gotten their defective pens down to 2% or less. If we pick it up from there, we will need several subgroup samples of data to establish the limits and process average for our chart. The *p*-chart construction process is very similar to the $\bar{x}$ and R charts discussed in the preceding section. For attributes data, the subgroup sample size should be larger. We need to have a sample size (n) large enough that we are likely to include the defectives. Let's use $n = 100$. We want the interval between sample groups wide enough that if trends develop, we will see them. If the factory makes 2,000 pens of this type per hour and we sample the first 100 after the hour, in an 8-hour day we can obtain eight samples. Three days of sampling will give us sufficient data to construct our *p*-chart. After 3 days of collecting data, we have the data shown in Figure 11–11. To that data, we'll apply the *p*-chart formulas shown in Figure 11–12.

Data	Subgroup	np	p	Data	Subgroup	np	p
8/11	1	1	.01	8/12	13	4	.04
8/11	2	2	.02	8/12	14	0	0
8/11	3	0	0	8/12	15	1	.01
8/11	4	0	0	8/12	16	3	.03
8/11	5	2	.02	8/13	17	1	.01
8/11	6	0	0	8/13	18	3	.03
8/11	7	3	.03	8/13	19	0	0
8/11	8	2	.02	8/13	20	2	.02
8/12	9	1	.01	8/13	21	4	.04
8/12	10	5	.05	8/13	22	1	.01
8/12	11	0	0	8/13	23	3	.03
8/12	12	2	.02	8/13	24	2	.02

np = number defective in subgroup   n = subgroup size = 100   k = # of subgroups

p = fraction defective $\bar{p}$ = process average

Figure 11–11
Collected Data for 3 Days

Figure 11–12
p-Chart Formulas

p = rejects in subgroup ÷ number inspected in subgroup
 = np ÷ n

$\bar{p}$ = total number of rejects ÷ total number inspected
 = Σnp ÷ Σn

$$UCL_p = \bar{p} + \frac{3\sqrt{\bar{p}(1-\bar{p})}}{\sqrt{n}}$$

$$LCL_p = \bar{p} - \frac{3\sqrt{\bar{p}(1-\bar{p})}}{\sqrt{n}}$$

Constructing the *p*-chart, we have several things to calculate: the fraction defective by subgroup (p), the process average ($\bar{p}$), and the control limits (UCL_p and LCL_p).

Fraction Defective by Subgroup (p)

The p values given in Figure 11–11 were derived by the formula $p = np \div n$. For example, for subgroup one, $np = 1$ (one pen was found defective from the first sample of 100 pens). Because p is the fraction defective,

$$p = 1 \div 100$$
$$= 0.01$$

For the second subgroup:

$$p = 2 \div 100$$
$$= 0.02$$

and so on.

Process Average ($\bar{p}$)

Calculate the process average by dividing the total number defective by the total number of pens in the subgroups:

$$\bar{p} = (np_1 + np_2 + \ldots + np_k) \div (n_1 + n_2 + \ldots n_k)$$
$$= 42 \div 2,400$$
$$= 0.0175$$

Control Limits (UCL_p and LCL_p)

Because this is the first time control limits have been calculated for the process (as shown in Figure 11–13), they should be considered *trial limits*. If we find that there

$$UCL_p = \bar{p} + \frac{3\sqrt{\bar{p}(1-\bar{p})}}{\sqrt{n}}$$
$$= .0175 + \frac{3\sqrt{.0175\,(1-.0175)}}{\sqrt{100}}$$
$$= .0175 + \frac{3\sqrt{.01719375}}{10}$$
$$= .0175 + \frac{3\times.1311}{10}$$
$$= .0175 + .0393$$
$$= .0568$$

$$LCL_p = \bar{p} - \frac{3\sqrt{\bar{p}(1-\bar{p})}}{\sqrt{n}}$$
$$= .0175 - \frac{3\sqrt{.0175\,(1-.0175)}}{\sqrt{100}}$$
$$= .0175 - \frac{3\sqrt{.01719375}}{10}$$
$$= .0175 - \frac{3\times.1311}{10}$$
$$= .0175 - .0393$$
$$= -.0218 \quad (\text{set at zero})$$

Figure 11–13
p-Chart Control Limit Calculations

are data points outside the limits, then we must identify the special causes and eliminate them. Then we can recalculate the limits without the special-cause data, similar to what we did in the series of Figures 11–8 through 11–10, but using the p-chart formulas.

In Figure 11–13, LCL_p is a negative number. In the real world, the fraction defective (p) cannot be negative, so we will set LCL_p at zero.

No further information is needed to construct the p-chart. The y-axis scale will have to be at least 0 to 0.06 or 0.07 because the $UCL_p = 0.0568$. The p values in Figure 11–11 do not exceed 0.05, although a larger fraction defective could occur in the future. Use the following steps to draw a control chart:

1. Label the x-axis and the y-axis.
2. Draw a dashed line representing UCL_p at 0.0568.
3. Draw a solid line representing the process average ($\bar{p}$) at 0.0175.
4. Plot the data points representing subgroup fraction defective (p).
5. Connect the points.

The p-chart (Figure 11–14) shows that there are no special causes affecting the process, so we can call it *in statistical control*.

Another Commonly Used Control Chart for Attributes Data

The c-Chart

The c-chart is used when the data is concerned with *the number of defects in a piece*— for example, the number of defects found in a tire or an appliance. In practice the data are collected by inspecting sample tires or toasters, whatever the product may be, on a scheduled basis, and each time logging the number of defects detected. Defects may also be logged by type (blemish, loose wire, and any other kind of defect noted), but the c-chart data are the simple sum of all the defects found in each sample piece. Remember, with the c-chart, a sample is one complete unit that may have multiple defect characteristics. The following example illustrates the development of a c-chart.

A manufacturer makes power supplies for the computer industry. Rework to correct defects has been a significant expense. The power supply market is very competitive, and for the firm to remain viable, defects and rework must be reduced. As a first step the company decides to develop a c-chart to help monitor the manufacturing process. To compile the initial data, the first power supply completed after the hour was

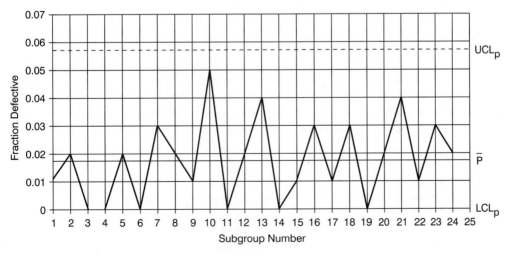

Figure 11–14
p-Chart

chosen as a sample and closely inspected. This was repeated each hour for 30 hours. Defects were recorded by type and totaled for each power supply sample. The data are shown in the table of Figure 11–16. From this data the initial c-chart was developed using the formulas of Figure 11–15.

Calculating the c-chart parameters from the data:

Total defects = 47
Number of samples = 30

$$\bar{c} = 47/30 = 1.56667$$

Largest c = 3
Smallest c = 0

$$\mathrm{UCL}_c = 1.56667 + 3\sqrt{1.56667}$$
$$= 4.32167$$
$$\mathrm{LCL}_c = 1.56667 - 3\sqrt{1.56667}$$
$$= -2.18833 \ (\text{Because this is negative, set to 0.})$$

The c-chart of Figure 11–17 is constructed from these data. Notice that all data points fell within the control limits, and there were no protracted runs of data points above or below the process average line, $\bar{c}$. The process was "in control" and ready for SPC. Now, as the operators continue to inspect a sample power supply each hour, data will immediately be plotted directly on the control chart, which of course will have to be lengthened horizontally to accept the new data. This is done with "pages" rather than physically lengthening the chart. Each new page represents a new control chart for the period chosen (week, month, etc.). However, the control limits and the average lines must remain in the same position until they are recalculated with new data. As process improvements are implemented and verified, recalculating the average and limits will be necessary. That is because when a process is really improved, it will have less natural variation. The original average and control limits would no longer reflect the process, and hence their continued use would invalidate the control chart.

The Control Chart as a Tool for Continuous Improvement

Control charts of all types are fundamental tools for continuous improvement. They provide alerts when special causes are at work in the process, and they prompt investigation and correction. When the initial special causes have been removed and the data stay between the control limits (within $\pm 3\sigma$), work can begin on process improvement. As process improvements are implemented, the control charts will either ratify the improvement or reveal that the anticipated results were not achieved. Whether the anticipated results were achieved is virtually impossible to know unless the process is under control. This is because there are special causes affecting the process; hence, one never knows whether the change made to the process was responsible for any subsequent shift in the data or if it was caused by something else entirely. However, once the process is in statistical control, any change you put into it can be linked directly to any shift in the subsequent data. You find out quickly what works and what doesn't. Keep the favorable changes, and discard the others.

Figure 11–15
c-Chart Formulas

$\bar{c}$ = total number of defects/number of samples
$\mathrm{UCL}_c = \bar{c} + 3\sqrt{\bar{c}}$
$\mathrm{LCL}_c = \bar{c} - 3\sqrt{\bar{c}}$

| | Defects by Type | | | | | | Number of Defects |
Sample	A	B	C	D	E	Other	
1			1	1			2
2		1	1		1		3
3	1						1
4							0
5			1				1
6		1		2			3
7			1			1	2
8			2	1			3
9							0
10	1		1			1	3
11		1					1
12				1	1		2
13	1	1					2
14			2				2
15				1		2	3
16							0
17		1					1
18			1	1			2
19							0
20			1				1
21							0
22						1	1
23		1			1		2
24			2		1		3
25	1						1
26			1				1
27			2			1	3
28							0
29				1		2	3
30			1				1
Sample: 1 Power Supply. Sample Rate: 1 Power Supply/hour							Total 47

Figure 11–16
c-Chart Data

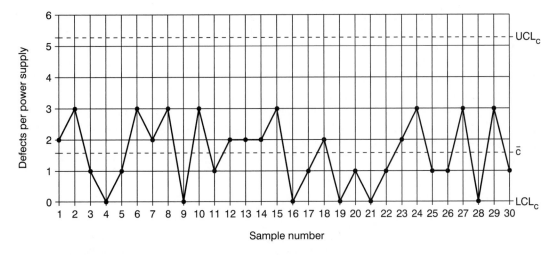

Figure 11–17
c-Chart: Power Supply Defects

As the process is refined and improved, it will be necessary to update the chart parameters. UCL, LCL, and process average will all shift, so you cannot continue to plot data on the original set of limits and process average. The results can look like the succession of charts in Figure 11–18.

An important thing to remember about control charts is that once they are established and the process is in statistical control, the charting does not stop. In fact, only then can the chart live up to its name, *control chart*. Having done the initial work of establishing limits and center lines and plotting initial data, and eliminating any special causes that were found, we have arrived at the starting point. Data will have to be continually collected from the process in the same way it was for the initial chart.

The plotting of these data must be done as it becomes available (in real time) so that the person managing the process will be alerted at the first sign of trouble in the process. Such trouble signals the need to stop the process and immediately investigate to determine what has changed. Whatever the problem is, it must be eliminated before the process is restarted. This is the essence of statistical process control. The control chart is the statistical device that enables SPC on the shop floor or in the office.

This discussion of control charts has illustrated only the $\bar{x}$ and R charts, the p-chart, and the c-chart. Figure 11–19 lists common control charts and their applications. The methods used in constructing the other charts are essentially the same as for the four we discussed in detail. Each chart type is intended for special application. You must determine which best fits your need.

Statistical Control versus Capability

It is important to understand the distinction between a process that is *in statistical control* and a process that is *capable*. Asking the question "Is our process in control?" is different from "Is our process capable?" The first relates to the absence of special causes in the process. If the process is in control, you know that 99.7% of the output will be within the $\pm 3\sigma$ limits. Even so, the process may not be capable of producing product that meets your customer's expectations.

Suppose you have a requirement for 500 shafts of 2-inch diameter with a tolerance of ± 0.02 inch. You already manufacture 2-inch diameter shafts in a stable process that is in control. The problem is that the process has control limits at 1.97 and 2.03 inches. The process is in control, but it is not capable of making the 500 shafts without a lot of scrap and the cost that goes with it. Sometimes it is possible to adjust the machines or procedures, but if that could have been done to tighten the limits, it already should have been. It is possible that a different machine is needed.

There are many variations on this theme. A process may be in control but not centered on the nominal specification of the product. With attributes data, you may want

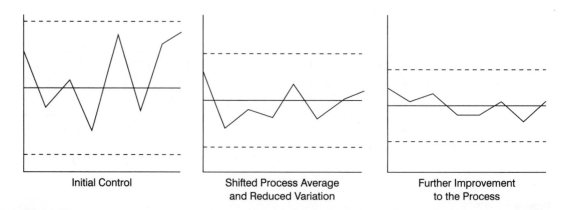

Initial Control   Shifted Process Average   Further Improvement
                               and Reduced Variation   to the Process

Figure 11–18
Succession of Control Charts

Data Category	Chart Type	Statistical Quantity	Application
Variables (measured values)	x bar-R ($\bar{x}$ & R)	Mean value and range	Charts dimensions and their precision weight, time, strength, and other measurable quantities. Example: Anything physically measurable.
	x tilde-R $\tilde{x}$ & r)	Median and range	Charts measurable quantities, similar tities, similar to $\bar{x}$ & R, but requires fewer calculations to plot. Example: Ditto above.
	x-Rs (also called x-chart)	Individual measured values	Used with long sample intervals: sub-grouping not possible. Example: Products made in batches like solutions, coatings, etc., or grouping too expensive (i.e., destructive testing). Histogram must be normal
Attributes (counted values)	p-chart	Percentage defective (also fraction defective)	Charts the number of defects in samples of varying size as a percentage or fraction. Example: Anywhere defects can be counted.
	np-chart (also pn)	Number of defective pieces	Charts the number of defective pieces in samples of fixed size. Example: As above, but in fixed size samples.
	c-chart	Number of defects	Charts the number of defects in a product (single piece) of fixed size (i.e., like products). Example: Specific assemblies or products (i.e., PC boards, tires, etc.).
	u-chart	Number of defects per unit area, time, length, etc.	Charts the number of defects in a product of varying size (i.e., unlike products). Example: Carpet (area), extrusions (length).

Figure 11–19
Common Control Charts and Their Applications

your in-control process to make 99.95% (1,999 out of 2,000) of its output acceptable, but it may be capable of making only 99.9% (1,998 of 2,000) acceptable. (Don't confuse that with $\pm 3\sigma$'s 99.73%: they are two different things.)

The series of charts in Figure 11–20 illustrates how *in statistical control* and *capable* are two different issues, but the control chart can clearly alert you to a capability problem. You must eliminate all special causes and the process must be in control before process capability can be established.

MANAGEMENT'S ROLE IN SPC

As in other aspects of total quality, management has a definite role to play in SPC. In the first place, as Dr. Deming has pointed out, only management can establish the

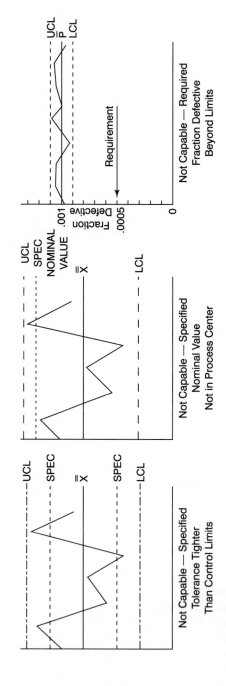

Figure 11–20
In Control and *Capable* Are Not the Same Thing

production quality level.[4] Second, SPC and the continuous improvement that results from it will transcend department lines, making it necessary for top management involvement. Third, budgets must be established and spent, something else that can be done only by management.

Commitment

As with every aspect of total quality, management commitment is an absolute necessity. SPC and continual improvement represent a new and different way of doing business, a new culture. No one in any organization, except its management, can edict such fundamental changes. One may ask why a production department cannot implement SPC on its own. The answer is that, providing management approves, it can. But the department will be prevented from reaping all the benefits that are possible if other departments are working to a different agenda. Suppose, for example, that through the use of SPC, a department has its processes under control and it is in the continuous improvements mode. Someone discovers that if an engineering change is made the product will be easier to assemble, reducing the chance for mistakes. This finding is presented to the engineering department. However, engineering management has budgetary constraints and chooses not to use its resources on what it sees as a production department problem. Is this a realistic situation? Yes, it is not only realistic, but very common. Unless there is a clear signal from top management that the production department's SPC program, with its continuous improvement initiative, is of vital interest, other departments will continue to address their own agendas. After all, each separate department knows what is important to the top management, and this is what they focus on because this is what affects their evaluations most. If SPC and continuous improvement are not perceived as priorities of top management, the department that implements SPC alone will be just that, alone.

Training

It is management's duty to establish the policies and procedures under which all employees work and to provide the necessary training to enable them to carry out those policies and procedures. The minimum management involvement relative to SPC training involves providing sufficient funding. More often, though, management will actually conduct some of the training. This is a good idea. Not only will management be better educated in the subject as a result of preparing to teach it, but employees will be more likely to get the message that SPC is important to management.

Involvement

When employees see management involved in an activity, they get a powerful message that the activity is important. Employees tend to align their efforts with the things they perceive as being important to management. If managers want their employees to give SPC a chance, they must demonstrate their commitment to it. This does not mean that managers should be on the floor taking and logging data, but they should make frequent appearances, learn about the process, probe, and insist on being kept informed.

A major part of SPC is the continuous improvement of processes. Deming pointed out that special causes of variation can be eliminated without management intervention. This is essentially true when it comes to correcting a problem. But when it comes to process improvement, management must be involved. Only management can spend money for new machines or authorize changes to the procedures and processes. Without management involvement, neither process nor product improvement will happen.

ROLE OF THE TOTAL QUALITY TOOLS

Some may consider it inappropriate to include tools other than control charts in a discussion of SPC. However, we take the broader view and include several tools:

- Pareto charts
- Cause-and-effect diagrams
- Stratification
- Check sheets
- Histograms
- Scatter diagrams
- Run charts and control charts
- Flowcharts
- Design of experiments

SPC does not start the moment a control chart is employed. Before SPC can be fully implemented, a lot of work must be done to eliminate the special causes of variation in the process concerned. Consequently, several quality tools will be used before it is time to develop and implement a control chart. When does SPC start? It starts when someone begins cleaning up the process. In the final analysis, this question is not that important, because the quality tools come into play either to support SPC or to be part of the SPC package, depending on the definition used.

With SPC, the total quality tools have a dual role. First, they help eliminate special causes from the process, so that the process can be brought under control. (Remember that a process that is in control has no special causes acting on it.) Only then can the control charts be developed for the process and the process monitored by the control charts. The second role comes into play when, from time to time, the control chart reveals a new special cause, or when the operator wants to improve a process that is in control. This is dealt with in the final section of this chapter, "Implementation and Deployment of SPC."

AUTHORITY OVER PROCESSES AND PRODUCTION

Operators who use SPC to keep track of their processes must have the authority to stop the production process when SPC tells them something is wrong. As long as the plots on the control chart vary about the process average but do not penetrate a control limit, the process is in control and is being influenced by the common causes of variation only. Once a penetration is made, or if the operator sees a run of several plots all on one side of the process average, he or she has good reason to believe that the process needs attention. The operator should be able to stop the process immediately.

A question that frequently arises in the early stages is "Can a line stoppage be justified in terms of cost and, possibly, schedule?" Toyota found early on that the value in stopping the line for any problem was absolute. Not only does it prevent waste and defective products, but having things at a standstill is a very powerful incentive for finding the cause of the problem and eliminating it—quickly. The word *eliminate* implies that the problem is corrected for good. In a less enlightened factory, the fear of a line stop is so great that the standard procedure is to apply a quick fix to get the line moving again. The emphasis is on keeping production rolling. With this approach, the problem will be investigated and corrected after the line is back in motion. The problem is this: the data that might have been available at that moment may have disappeared by the time someone looks for it, and the trail to the root cause may be lost. The odds are good that the same event will recur later on with similar disruption and impact on quality. It will keep recurring until the cause is finally discovered and eliminated. Under

SPC and total quality, the emphasis is not on maintaining production no matter what but on eliminating any cause of substandard quality the moment it comes up.

Attempting SPC without giving process-stopping authority to the operator is a serious error. Harry Truman once said that war was too important to leave in the hands of the generals. We believe that line stoppages are too important to leave in the hands of management. Give the authority to the operator, and the underlying problems will be eliminated.

IMPLEMENTATION AND DEPLOYMENT OF SPC

Going into SPC is not something to be taken lightly or approached in a halfhearted manner. It requires the time and commitment of key personnel. It involves training and the expenses that go with it. It may even involve hiring one or more new people with specialized skills. There may be expenses for consultants to help get the organization started and checked out in SPC. The organization may have to invest in some new tools/tooling, if what is already on hand turns out to be inadequate. But the single most important issue that must be faced when implementing SPC is the culture change that is implicit in using SPC.

Up to this point, the organization has relied on the quality department to assure the quality of products. With SPC, the process operator must be the one who assures product quality, and the quality department must step aside, taking on a significantly different role. Before, if operators could do the assembly steps necessary for their processes, that was adequate. Now their scope of activity must be expanded into new areas, and they must be helped to develop the skills needed to cope with the new requirements. Supervisors and middle managers must give operators the latitude and freedom required to perform the new functions effectively. This sounds easier than it actually is. Many people find it difficult to adjust and adapt.

In addition, when SPC is used, functions that were formerly done by individuals will increasingly be performed by collaborative teams. Employees learn that solving problems, using the quality tools, even defining their own processes are best done by teams of people who bring to the table an array of skills, knowledge, and viewpoints that would be impossible for the individual. Interpreting the control charts, finding root causes of any detected special cause events, and developing ways to actually improve processes are examples of new tasks that come with SPC. All require team activity.

There is no single right way to implement SPC. What is presented here is a general road map for implementation, covering the major steps in the chronological order in which they should be introduced. The detail behind each of the steps must be worked out for each unique application. An SPC implementation is one area in which the retention of an expert consultant has merit. For SPC to provide any benefit, the program must be statistically valid, and it takes an expert to know whether it is or not.

Figure 11–21 summarizes the steps involved in implementing SPC. The implementation steps are divided into three phases: preparation, planning, and execution.

The Preparation Phase

The preparation phase for SPC includes three steps.

Step 1: Commit to SPC

Any endeavor that requires money, human resources, changing the organization's culture, hiring employees with new skills, or retaining consultants is something to which top management must be committed. The department that forges ahead without that commitment may find itself cut off in mid-effort, a situation worse than not having started at all.

Phase	Responsibility	Action
Preparation	Top management Top management Consultant or in-house expert	(1) Commitment to SPC (2) Organize SPC Committee (3) Train SPC Committee
Planning	SPC Committee assisted by consultant or expert Consultant or in-house expert QA Management	(4) Set SPC Objectives (5) Identify Target Processes (6) Train Appropriate Operators and Support Personnel (7) Assure Repeatability and Reproduceability of Instruments and Methods (8) Delegate Responsibility for Operators to Play Key Role
Execution	SPC Committee, operator, suppliers, customers Operator w/ expert assist. Consultant or in-house expert Operator Operator w/ expert assist. Operator SPC Committee and management Operator w/ assist. as required All	(9) Flow Chart the Process (10) Eliminate the Special Causes of Variation (11) Develop Control Chart(s) (12) Collect and Plot SPC Data (13) Determine Process Capability* (14) Respond to Trends and Out-of-Limits Data (15) Track SPC Data (16) Eliminate Root Causes of any Special Causes of Variation (17) Continuously Improve the Process (narrow the limits)

Figure 11-21

The SPC Implementation Road Map

* If the process is not capable of meeting requirements, it must be changed or replaced; go back to step 9.

Step 2: Form an SPC Committee

SPC can take a lot of time, especially at first when employees are getting acquainted with it and are getting the processes on-line. Unlike total quality, however, SPC can be delegated to a cross-functional team that is tasked to oversee implementation and execution. The SPC team leader need not be the resident expert (at the beginning), but a statistics expert must be included on the team and that person must be heard. A typical team will be composed of representatives from manufacturing, quality assurance, engineering, finance, and statistics. In a manufacturing plant, the manufacturing member should be the team leader. The function of the team will be to plan and organize the implementation for its unique application, to provide training for the operators, and to monitor and guide the execution phase. Forming the SPC committee is top management's responsibility.

Step 3: Train the SPC Committee

The newly formed SPC committee must receive basic training before its work starts. In a typical situation, the committee members will have had little or no practical experience with statistics. The training must be done by an expert. It is possible to send employees to training courses or bring the expert to the company. At the conclusion of the training period, the members will not have become experts, but they will know enough to set objectives and to determine which processes should be targeted first. At this point, continued help from a statistics expert remains critical.

The Planning Phase

The planning phase includes the next five steps.

Step 4: Set SPC Objectives

The SPC committee should set objectives for the program. What do we hope to gain from SPC? How will we measure success (at the balance sheet's bottom line, customer feedback, reduction in scrap, lower cost of quality, or perhaps all of these)? If the team waits until the SPC machinery is in place and producing data to decide what gains are expected, consensus may never be reached on how well or how poorly it is working. Set the objectives. Measure against them. As with all objectives, they should be reviewed from time to time to make sure they are still valid and meaningful. Objectives may be added, eliminated, or changed, but they must be in place and understood by all.

Step 5: Identify Target Processes

It is not feasible to attempt to apply SPC to all processes at once. The people involved in designing the SPC application, collecting data and interpreting its meaning, getting the processes under control, and plotting and evaluating control chart data will be in a learning mode for the first several weeks. For that reason, it is important to select just a few pilot processes for the initial implementation. These should be processes that are well understood and that promise to be relatively easy to bring under control. They should also be important processes, ones that have meaning rather than something trivial. The key point to remember is this: Select initial processes from among those that stand the best chance of quick success. With some initial successes under its belt, the organization can go on with confidence to the processes that are the most critical.

Consideration should also be given to the flow of processes one to another. For example, if there is a production line with four processes, it makes sense to implement SPC in the order of production flow. Trying to introduce it at the end or in the middle of the four processes may prove difficult. If the first three processes feed their defects into the final process, it will be impossible to eliminate the special causes of variation of the fourth. On the other hand, by starting at the beginning of the flow, putting process number one under control may eliminate one or more of the special causes affecting processes two through four. The idea is to start implementation at the front of a series of processes, not at the back. Selection of the target processes should be done by the SPC committee, with comprehensive, open communication with the process operators.

Step 6: Train Appropriate Operators and Teams

The operators and teams who will be directly involved with the collection, plotting, and interpretation of SPC data, and those who will be involved in getting the targeted processes under control, will require training in the use of quality tools and in flowcharting. Some processes may require the use of design of experiments (DOX). If this is the case, the help of a specialist, both to provide training and to assist with the DOX process, may be needed. Training given at this point must make clear the significance and the objectives of the work to be undertaken. Participants will be the process operators and the engineers and quality specialists who support them. Only the employees who will be involved in the initial SPC projects should be included in the first class. As SPC is spread throughout the plant, it will be necessary to train other operators and teams and their support personnel. But by delaying training until it is time to expand beyond the initial processes, the advantage of just-in-time training will be gained. In addition, you will be able to capitalize on lessons learned from the initial projects. The training needed can typically be accommodated in a 1- or 2-day session.

Step 7: Assure Repeatability and Reproducibility (R&R) of Gauges and Methods

All measuring instruments, from simple calipers and micrometers to coordinate measuring machines, must be calibrated and certified for acceptable R&R performance. For SPC to work, the measured data plotted on the control charts must be reliable. A gauge that cannot repeat the same measurement with the same operator consistently, or one that is so difficult or idiosyncratic in its use that no two operators can obtain the same data, will not work in an SPC environment. The particular application will determine the range of variability that is acceptable in measuring instruments. It must then be verified that each instrument to be used is capable — and that all the people who will be using the instrument are adequately trained in its use. This must be done before step 10.

Step 8: Delegate Responsibility for Operators to Play a Key Role

As the last step in the planning phase, just before SPC execution is to begin, management should delegate to the process operators responsibility for maintaining the SPC control charts, collecting and plotting the data, and taking appropriate action. Let the operators know that these functions are theirs, but make certain everyone else knows it too.

The Execution Phase

The execution phase includes nine steps.

Step 9: Flowchart the Process

The first step in the SPC execution phase, taking the broad perspective regarding SPC's boundaries, is flowcharting or characterizing the process to which SPC will be applied. Only when a graphic representation of the entire process exists—including its inputs, outputs, and all the steps between—can the process be fully understood. Invariably, flowcharting will reveal process features or factors that were not known to everyone. After the flowchart has been completed and everyone agrees that it represents the way the process actually works, a large version should be produced on poster board and permanently placed in open view at the process location. It will provide invaluable information and may even suggest process improvements later on. Members of the SPC team should help, but the development of the process flowcharts should be the responsibility of special teams composed of the process operators, their internal suppliers and customers, and appropriate support members. (Support personnel may include engineers, materials specialists, financial specialists, etc., as needed.)

Step 10: Eliminate the Special Causes of Variation

Now that participants understand the process, it is time to identify and eliminate the special causes of variation. This is best begun through the use of the cause-and-effect diagram, which was discussed in Chapter 9 as one of the seven total quality tools. The cause-and-effect diagram will list all of the factors (causes) that might impact the output in a particular way (effect). Then by applying the other tools, such as Pareto charts, histograms, and stratification, the special causes can be identified and eliminated. Until the special causes that are working on the process are eliminated, the next steps will be difficult or impossible. Elimination of the special causes should be a team effort among the process operators, internal process suppliers and customers, engineers, and quality assurance personnel, with additional help from other departments as required. For example, if materials are a factor, the purchasing department might become involved. Be sure to keep the operators at the center of the activity, as this will give them ownership as well as valuable experience.

Step 11: Develop Control Charts

With the absence of special causes, it is now possible to observe the process unencumbered by external factors. The statistics expert, or consultant, can now help develop the appropriate control charts and calculate valid upper and lower control limits and process averages. Selection of the control chart type will be determined by the kind of data to be used. (See Figure 11–19.)

Step 12: Collect and Plot SPC Data; Monitor

With the special causes removed, and with the process running without tweaking, the process operator takes the sample data (as specified by the statistics expert) and plots it on the control chart at regular intervals. The operator carefully observes the location of the plots, knowing that they should be inside the control limits, with the pattern varying randomly about the process average if the process is in control.

Step 13: Determine Process Capability

Before going further, it is important to determine whether the process is capable of doing what is expected of it. For example, if the process output is to be metal parts with a specified length of between 5.999 and 6.001 inches, but the process turns out as many pieces outside those dimensions as it does within, the process is not capable. The process is capable if its frequency distribution is a bell-shaped curve centered on the specification average, in this case 6 inches, and with the $\pm 3\sigma$ spread coincident with, or narrower than, the specification limits. With the bell curve centered on the specification average, and the specification limits coincident with the $\pm 3\sigma$ spread of the bell curve, we could expect 3 nonconforming parts out of 1,000. If the specification limits are inside the $\pm 3\sigma$ spread, then the defect rate would be higher; if they are outside (the bell curve is narrower than the limits), the defect rate would be lower.

Two methods exist for determining process capability. The first assumes that the bell curve is centered on the specification average and is called C_p. The second does not assume alignment of the process average and the specification average, and it is called C_{pk}. Figure 11–22 explains the procedures for calculating these capability indices. As we have already learned, it is possible to have a process that is in control and still not capable of meeting the customer's specifications. When this is the case, it is up to

Process Capability may be calculated in two ways. The first assumes that the process average is centered on the specification average, and is denoted as C_p, the process Capability Index.

$$C_p = \frac{USL\text{-}LSL}{6\hat{\sigma}}$$

Where: USL = Upper Specification Limit
LSL = Lower Specification Limit

$\hat{\sigma}$ = Estimated Process Average

$$\hat{\sigma} = \frac{\bar{R}}{d_2}$$

Where: d_2 **is a constant** (see table below)

$\bar{R}$ = the Process Average Range

The second method is used when the process average is not assumed to be coincident with the specification average, and is denoted as C_{pk}.

$$C_{pk} = \frac{USL\text{-}\bar{\bar{X}}}{3\hat{\sigma}}$$

and

$$C_{pk} = \frac{\bar{\bar{X}}\text{-}LSL}{3\hat{\sigma}}$$

Where: $\bar{\bar{X}}$ = the Process Average

C_{pk} is taken as the *smaller of the two values.*

For either case, a Capability Index of:

=1 means that the specification limits and average are coincident with the process ± 3σ limits and process average.

<1 means that the specification is tighter than the process spread. The defect rate will be greater than 3 parts per 1000.

>1 means that the process spread is tighter than the specification limits. The defect rate will be less than 3 parts per 1000.

NOTE: 1.33 IS THE PREFERRED MINIMUM CAPABILITY INDEX

Table for d_2 Values

# Observations in subgroup	2	3	4	5	6	7	8	9
d_2	1.128	1.693	2.059	2.326	2.534	2.704	2.847	2.970

Figure 11–22
Process Capability Calculation

198

management to replace or upgrade the process capability, which may require the purchase of new equipment.

Step 14: Respond to Trends and Out-of-Limits Data

As data are plotted, the operator must respond to any penetration of the control limits or to any run of data above or below the process average line. Either of these is an indication that something is wrong within the process or that some external factor (a special cause) has influenced the process. With experience, operators may be able to handle many of these situations on their own, but when they cannot, it is important that they summon help immediately. The process should be stopped until the cause is identified and removed. This is one of the most important functions of SPC—letting the operator know there is a problem early enough to prevent the production of defective products that must be scrapped or reworked. The only way to respond in such cases is by immediately eliminating the problem. This is another application for team (usually ad hoc) participation.

Step 15: Track SPC Data

The SPC committee and management should pay close attention to the SPC data that are generated on the production floor. Doing so will give them an accurate picture of their production capability, the quality of their processes, trends that may develop, and where they should concentrate resources for improvement. A secondary benefit of displaying this level of interest in SPC is that the operators and their support functions will know that management is truly interested in the program, and they will give it the attention and care appropriate to a high-visibility initiative.

Step 16: Eliminate the Root Causes of Any New Special Causes of Variation

From time to time, new special causes will come up, even in processes that have long been in control. When this happens, the operator will know it because the SPC data will go out of limits or skew to one side or the other of the control chart center line. It is important that the root causes of these special causes be eliminated to prevent their recurrence. For example, if the purchasing department placed an order for the next shipment of raw material from a different vendor because its price was cheaper than the current supplier, it is possible that the material coming from the new source might react differently in the process, shifting the process average one way or the other. The root cause may not be the new material. If you scrap it, purchasing is more than likely going to order the replacement material from the same low bidder, and the problem will probably recur. It would seem that the root cause of the problem is purchasing's tendency to order from the cheapest source. Eliminating this root cause may require a management-approved procedure mandating the use of preferred suppliers. At the very least, there should be an ironclad agreement that purchasing would not order materials from a new supplier without having the material certified by quality assurance and manufacturing personnel. This is a case where the operator initiates the action, a team may identify the root cause, and management involvement may be required to eliminate the problem. This is the way the process is meant to proceed. Wherever the help must come from, it has to be readily available.

Step 17: Continuous Improvement: Narrow the Limits

With the process under control and the special causes eliminated, continuous improvement can be implemented. What this means is that the process average should be centered on the specification average, if that is not already the case, and more frequently, it means the narrowing of the $\pm 3\sigma$ limits (see Figure 11–23). Both of these improvements—centering the process on the specification average and narrowing the limits—will result in fewer parts failing to meet the specifications. Scrappage will be reduced,

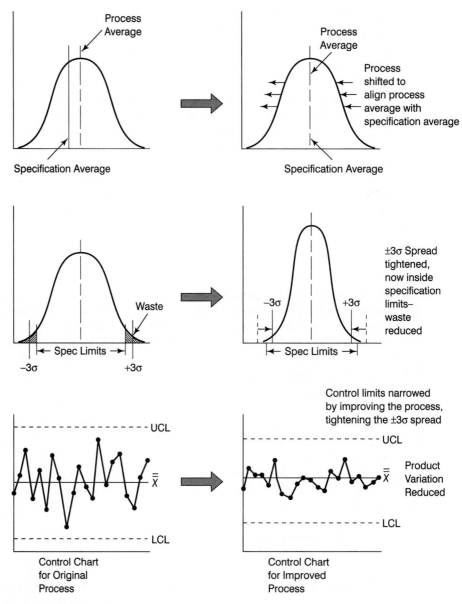

Figure 11–23
Process Improvement

the process will become more robust, quality will improve, and costs will decrease. The key, of course, is finding ways to improve the processes, but with SPC, one has the understanding of the processes necessary to see and comprehend the problems. Only then can real improvement follow.

INHIBITORS OF SPC

As in the case of implementing just-in-time (JIT) and benchmarking, a number of factors can inhibit the implementation of SPC. With SPC there is not usually the kind of philosophical resistance that is common with JIT and benchmarking. However, it is true even with SPC that there must be a management commitment, because there will be startup costs associated with implementation. The most common inhibitor of SPC is lack of resources.

Capability in Statistics

Many organizations do not have the in-house expertise in statistics that is necessary for SPC. As SPC is being introduced and decisions are made on where to sample, how much to sample, what kinds of control charts to use, and so forth, a good statistician is necessary to assure the validity of the program. If the organization does not employ such an expert, it should either hire one or retain the services of a consultant for the early phases of the SPC implementation.

The danger inherent in not having statistical expertise is developing an SPC program that is statistically invalid—a fact that can easily escape nonstatisticians. The organization will count on SPC to control processes when, in fact, it cannot. A flawed SPC implementation may send messages that make the process control situation worse than it was before. It is important that the initial design of the SPC program be valid. This requires someone with more than a passing knowledge of statistics. If there are any doubts, get help.

Misdirected Responsibility for SPC

Too many companies make the decision to use SPC but then turn it over to the statisticians or the quality assurance department. The value of those departments should not be minimized, but the owner of the process in question should be the person responsible for SPC. This person is the one who can make best use of SPC, and there will be no question about the validity of the data because he or she is the one collecting it. The process operators will require help from the statistician and others from time to time, but they are the appropriate owners of SPC for their processes.

When someone else is responsible for SPC (meaning the collection and logging of data and making corrections, stopping out-of-control processes, and getting them fixed), process operators see the entire SPC program as just another check on them, and they are very uncomfortable with it. Management tends to see it the same way, but from their particular perspective—a means of checking up on the operators. Nothing good will come from such a relationship.

Neither is ownership by the statistician the appropriate answer. If the statistician owns SPC, he or she is more apt to find fascination in the numbers themselves than in what they mean in terms of quality. Even if statisticians are tuned in to the objective, the operator will see them and their SPC charts as just another intrusion.

If operators have the responsibility for SPC, they will become familiar with the tasks involved and will see it as a means to help them get the most out of their processes. This is the payoff. All of the others need to observe and review, assisting when needed, but never usurping the operator's ownership.

Failure to Understand the Target Process

Unless a process has been flowcharted recently and characterized, the odds are good that the people designing the SPC system for it do not know how the process actually works. Most processes have evolved over many years, changing now and then to meet requirements of the market or the desires of management or operators. Few are adequately documented. People are usually astonished when flowcharts reveal the complexity of processes they thought to be straightforward. A good SPC system cannot be designed for a process that isn't fully understood.

Failure to Have Processes under Control

Before SPC can be effective, any special causes of variation must be removed. This was discussed earlier, but it is appropriate to mention it here again as an inhibitor of SPC.

Remember, by definition, a process is not in control if any special cause is working on it. The use of control charts assumes an in-control process. Their use will set off visual alarms whenever a new special cause is introduced. But the real work of process improvement can come about only when nothing but the common causes are active. This is why SPC is so powerful. It will show when common causes are the only causes of variation so that improvements to the fundamental process can be made. Will special causes still come up from time to time? Certainly, but this is different from trying to control a process with special causes constantly present, masking both the common causes and each other.

Inadequate Training and Discipline

Everyone who will be involved in the SPC program must be trained, not only in data acquisition, plotting, and interpretation of control charts but also in the use of the seven tools. Not everyone needs to be a statistics expert, but all need to know enough so that with a statistician's help, the program can be designed and operated.

Training should teach that SPC and tweaking do not make a good pair. If tweaking (frequent minor adjustments to one or more process factors) is permitted, SPC data will be meaningless. The process may appear to be more stable than it is if the person doing the tweaking is an expert, or it may show more variation than if left alone. It must be understood that operators and engineers alike are to let the process run essentially hands-off until an out-of-limits condition is detected. Variation between the limits is not to be tweaked out. The only acceptable means of reducing the variation is a real process improvement that will narrow the limits permanently.

Measurement Repeatability and Reproducibility

SPC data are the result of measurement or count. In the case of the variables data (the measurements), the data becomes meaningless when the measurements are not repeatable. For example, a worn instrument or a gauge with insufficient precision and resolution might yield measurements over an unacceptably wide range when measuring the same object repeatedly. This is not satisfactory. The data taken from all measurements must be accurate to the degree specified, and repeatable, or there is no point in recording it.

Nothing should be taken for granted. Before any gauge is used for SPC, it should be calibrated and its repeatability certified. It is also important that different operators obtain the same readings. This is known as *reproducibility*. Before getting them involved with SPC, certify all gauges and train all operators.

Low Production Rates

Although it is more convenient to implement SPC with processes that have continuous flow, or high rates of product output, it is by no means impossible to apply SPC to low-rate production of the type that is often found in a job shop setting. In a factory that produces several hundred printed circuit boards per day, sampling schemes are relatively easy. A job shop might produce only a few boards in a day, often with gaps between production days. Sampling there must be done differently. Low-rate production provides an opportunity for taking a 100% sample. It is possible to take a sample from every board. In such an application, a computer-generated random x-y table can designate a specific small area of a board for inspection of solder joints or other attributes. From that, a number representing the fraction defective may be developed. Control charts are easily constructed for fraction-defective data. Low production rates are not a good excuse for avoiding SPC.

══════════ ENDNOTES ══════════

1. Mary Walton, *The Deming Management Method* (New York: Putnam, 1986), 12.
2. Philip B. Crosby, *Quality Is Free: The Art of Making Quality Certain* (New York: McGraw-Hill, 1979), 68.
3. Gregory B. Hutchins, *Introduction to Quality Control, Assurance, and Management* (New York: Macmillan, 1991), 162.
4. Nancy R. Mann, *The Keys to Excellence: The Story of the Deming Philosophy* (Los Angeles: Prestwick, 1989), 21.

CHAPTER TWELVE

Continuous Improvement

One of the most fundamental elements of total quality is continuous improvement. The concept applies to processes and the people who operate them. It also applies to products. However, a fundamental total quality philosophy is that the best way to improve a product is to continually improve the processes by which it is made. This chapter provides the information needed to make continual workplace improvements.

RATIONALE FOR CONTINUOUS IMPROVEMENT

Continuous improvement is fundamental to success in the global marketplace. Companies that are just maintaining the status quo in such key areas as quality, new product development, the adoption of new technologies, and process performance are like a runner who is standing still in a race. Competing in the global marketplace is like competing in the Olympics. Last year's records are sure to be broken this year. Athletes who don't improve continually are not likely to remain long in the winner's circle. The same is true of companies that must compete globally.

Customer needs are not static. They change continually. A special product feature that is considered innovative today will be considered just routine tomorrow. A product cost that is considered a bargain today will be too high to compete tomorrow. A good case in point in this regard is the ever-falling price for each new feature introduced in the personal computer. The only way a company can hope to compete in the modern marketplace is to improve continually.

MANAGEMENT'S ROLE IN CONTINUOUS IMPROVEMENT

In his book *Juran on Leadership for Quality*, Joseph Juran writes:

> The picture of a company reaping big rewards through quality improvement is incomplete unless it includes some realities that have been unwelcome to most upper managers. Chief among these realities is the fact that the upper managers must participate personally and extensively in the effort. It is not enough to establish policies, create awareness, and then leave all else to subordinates. That has been tried, over and over again, with disappointing results.[1]

Management can play the necessary leadership role—and that essentially is its role—in continuous improvement by doing the following:

- Establishing an organization-wide quality council and serving on it
- Working with the quality council to establish specific quality improvement goals with timetables and target dates
- Providing the necessary moral and physical support. Moral support manifests itself as commitment. Physical support comes in the form of the resources needed to accomplish the quality improvement objectives.
- Scheduling periodic progress reviews and giving recognition where it is deserved
- Building continuous quality improvement into the regular reward system, including promotions and pay increases

ESSENTIAL IMPROVEMENT ACTIVITIES

Continuous improvement is not about solving isolated problems as they occur. Such an approach is viewed as putting out fires by advocates of total quality. Solving a problem without correcting the fault that caused it—in other words, simply putting out the fire—just means the problem will occur again. Quality expert Peter R. Scholtes recommends the following five activities that he sees as crucial to continuous improvement.[2]

- *Maintain communication.* Communication is essential to continuous improvement. This cannot be overemphasized. Communication within improvement teams and between teams is a must. It is important to share information before, during, and after attempting to make improvements. All people involved as well as any person or unit that might be impacted by a planned improvement should know what is being done, why, and how it might affect them.
- *Correct obvious problems.* Often process problems are not obvious, and a great deal of study is required to isolate them and find solutions. This is the typical case, and it is why the scientific approach is so important in a total quality setting. However, sometimes a problem with a process will be obvious. In such cases, the problem should be corrected immediately. Spending days studying a problem for which the solution is obvious just so that the scientific approach is used will result in $10 solutions to $0.10 problems.
- *Look upstream.* Look for causes, not symptoms. This is a difficult point to make with people who are used to taking a cursory glance at a situation and putting out the fire as quickly as possible without taking the time to determine what caused it.
- *Document problems and progress.* Take the time to write it down. It is not uncommon for an organization to continue solving the same problem over and over because nobody took the time to document the problems that have been dealt with and how they were solved. A fundamental rule for any improvement project team is "document, document, document."

- *Monitor changes.*   Regardless of how well studied a problem is, the solution eventually put in place may not solve it or may only partially solve it. For this reason, it is important to monitor the performance of a process after changes have been implemented. It is also important to ensure that pride of ownership on the part of those who recommended the changes does not interfere with objective monitoring of the changes. These activities are essential regardless of how the improvement effort is structured.

STRUCTURE FOR QUALITY IMPROVEMENT

Quality improvement doesn't just happen. It must be undertaken in a systematic, step-by-step manner. For an organization to make continuous improvements, it must be structured appropriately. Quality pioneer Joseph Juran calls this "mobilizing for quality improvement."[3] It involves the following steps:[4]

- *Establish a quality council.*   The quality council has overall responsibility for continuous improvement. According to Juran, "The basic responsibility of this council is to launch, coordinate, and 'institutionalize' annual quality improvement."[5] It is essential that the membership include executive-level decision makers.

- *Develop a statement of responsibilities.*   All members of the quality council, as well as employees who are not currently members, must understand the council's responsibilities. One of the first priorities of the council is to develop and distribute a statement of responsibilities bearing the signature of the organization's CEO. Responsibilities that should be stated include the following: (a) formulating policy as it relates to quality; (b) setting the benchmarks and dimensions (cost of poor quality, etc.); (c) establishing the team and project selection processes; (d) providing the necessary resources (training, time away from job duties to serve on a project team, and so on); (e) implementing the project; (f) establishing quality measures for monitoring progress and undertaking monitoring efforts; and (g) implementing an appropriate reward and recognition program.

- *Establishing the necessary infrastructure.*   The quality council constitutes the foundation of an organization's quality effort. However, there is more to the quality infrastructure than just the council. The remainder of the quality infrastructure consists of subcommittees of the council that are assigned responsibility for specific duties, project improvement teams, quality improvement managers, a quality training program, and a structured improvement process.

THE SCIENTIFIC APPROACH

The scientific approach is one of the fundamental concepts that separates the total quality approach from other ways of doing business. Scholtes describes the scientific approach as "making decisions based on data, looking for root causes of problems, and seeking permanent solutions instead of relying on quick fixes."[6]

Scholtes developed the following four strategies for putting the scientific approach to work in a total quality setting:[7]

- Collect meaningful data.
- Identify root causes of problems.
- Develop appropriate solutions.
- Plan and make changes.

Collect Meaningful Data

Meaningful data are free from errors of measurement or procedure, and they have direct application to the issue in question.[8] It is not uncommon for an organization or a unit within it to collect meaningless data or to make a procedural error that results in the collection of erroneous data. In fact, in the age of computers, this is quite common. Decisions based on meaningless or erroneous data are bound to lead to failure. Before collecting data, decide exactly what is needed, how it can best be collected, where the data exists, how it will be measured, and how you will know the data are accurate.

Identify Root Causes of Problems

The strategy of identifying root causes is emphasized throughout this book.[9] Too many resources are wasted by organizations attempting to solve symptoms rather than problems. The total quality tools are helpful in separating problems from causes.

Develop Appropriate Solutions

With the scientific approach, solutions are not assumed.[10] Collect the relevant data, make sure it is accurate, identify root causes, and then develop a solution that is appropriate. Too many teams and too many people begin with "I know what the problem is. All we have to do to solve it is. . . . " When the scientific approach is applied, the problem identified is often much different from what would have been suspected if acting on a hunch or intuition. Correspondingly, the solution is also different.

Plan and Make Changes

Too many decision makers use what is sometimes called the "Ready, fire, aim" approach rather than engaging in careful, deliberate planning.[11] Planning forces you to look ahead, anticipate needs and what resources will be available to satisfy them, and anticipate problems and consider how they should be handled.

Much of the scientific approach has to do with establishing reliable performance indicators and using them to measure actual performance. In his book *Total Manufacturing Management*, Giorgio Merli lists the following as examples of useful performance indicators:[12]

- Number of errors or defects
- Number of or level of need for repetitions of work tasks
- Efficiency indicators (units per hour, items per person)
- Number of delays
- Duration of a given procedure or activity
- Response time or cycle
- Usability/cost ratio
- Amount of overtime required
- Changes in workload
- Vulnerability of the system
- Level of criticalness
- Level of standardization
- Number of unfinished documents

This is not a complete list. Many other indicators could be added. Those actually used vary widely from organization to organization. However, such indicators, regardless of which are actually used, are an important aspect of the scientific approach.

IDENTIFICATION OF IMPROVEMENT NEEDS

Even the most competitive, most successful organizations have limited resources. Therefore, it is important to optimize those resources and use them in ways that will yield the most benefit. One of the ways to do this is to carefully select the areas of improvement to which time, energy, and other resources will be devoted. If there are 10 processes that might be improved on, which will yield the most benefit if improved? These are the processes that should be worked on first.

Scholtes recommends the following four strategies for identifying improvement needs:[13]

- *Apply multivoting.*   Multivoting involves using brainstorming to develop a list of potential improvement projects. Team members vote several times—hence the name—to decide which project or projects to work on first. Suppose the original list contains 15 potential projects. Team members vote and cut the list to 10. They vote again and cut it to five. The next vote cuts the list to three and so on until only one or two projects remain. These are the first projects that will be undertaken.
- *Identify customer needs.*   An excellent way to identify an improvement project is to give the customer a voice in the process. Identify pressing customer needs and use them as projects for improvement.
- *Study the use of time.*   A good way to identify an improvement project is to study how employees spend their time. Is an excessive amount of time devoted to a given process, problem, or work situation? This could signal a problem. If so, study it carefully to determine the root causes.
- *Localize problems.*   Localizing a problem means pinpointing specifically where it happens, when it happens, and how often. It is important to localize a problem before trying to solve it. Problems tend to be like roof leaks in that they often show up at a location far removed from the source.

Big Q versus Little Q in Project Selection

Total quality pioneer Joseph M. Juran emphasizes the need to understand the difference between what he calls "Big Q" and "Little Q" when selecting improvement projects.[14] Little Q is a much more specific, more focused, more restricted aspect of quality. For example, deciding to improve a given feature of a specific product falls under the heading of Little Q. Improving all of the processes and people associated with producing this product falls under the heading of Big Q. Little Q sees the customer as being the person who buys the product. Big Q sees customers as all people involved, internal and external.[15] It is important to understand the difference when selecting improvement projects because Big Q-oriented projects will yield more and better long-term results than Little Q projects.

DEVELOPMENT OF IMPROVEMENT PLANS

After a project has been selected, a project improvement team is established. The team should consist of representatives from the units most closely associated with the problem in question. It must include a representative from every unit that will have to be involved in carrying out improvement strategies. The project improvement team should begin by developing an improvement plan. This is to make sure the team does not take the "Ready, fire, aim" approach mentioned earlier.

The first step is to develop a mission statement for the team. This statement should clearly define the team's purpose and should be approved by the organization's governing board for quality (executive committee, quality council, or whatever the group is). After this has been accomplished, the plan can be developed. Scholtes recommends the following five stages for developing the plan:

1. *Understand the process.*   Before attempting to improve a process, make sure every team member thoroughly understands it. How does it work? What is it supposed to do? What are the best practices known pertaining to the process? The team should ask these questions and pursue the answers together. This will give all team members a common understanding, eliminate ambiguity and inconsistencies, and point out any

obvious problems that must be dealt with before proceeding to the next stage of planning.

2. *Eliminate errors.*   In analyzing the process, the team may identify obvious errors that can be quickly eliminated. Such errors should be eliminated before proceeding to the next stage. This stage is sometimes referred to as "error-proofing" the process.

3. *Remove slack.*   This stage involves analyzing all of the steps in the process to determine whether they serve any purpose and, if so, what purpose they serve. In any organization, processes exist that have grown over the years with people continuing to follow them without giving any thought to why things are done a certain way, whether they could be done better another way, or whether they need to be done at all. Few processes cannot be streamlined.

4. *Reduce variation.*   Variation in a process results from either common causes or special causes. Common causes result in slight variations and are almost always present. Special causes typically result in greater variations in performance and are not always present. Strategies for identifying and eliminating sources of variation are discussed in the next section of this chapter.

5. *Plan for continuous improvement.*   By the time this step has been reached, the process in question should be in good shape. The key now is to incorporate the types of improvements made on a continual basis so that continuous improvement becomes a normal part of doing business. The Plan-Do-Check-Act cycle applies here. With this cycle, each time a problem or potential improvement is identified, an improvement plan is developed (Plan), implemented (Do), monitored (Check), and refined as needed (Act).

COMMON IMPROVEMENT STRATEGIES

Numerous different processes are used in business and industry; consequently, there is no single road map to follow when improving processes. However, a number of standard strategies can be used as a menu from which improvement strategies can be selected as appropriate. These strategies are explained in the following sections.[16]

Describe the Process

The strategy of describing the process is used to make sure that everyone involved in improving a process has a detailed knowledge of the process. Usually this requires some investigation and study. The steps involved are as follows:

1. Establish boundaries for the process.
2. Flowchart the process.
3. Make a diagram of how the work flows.
4. Verify your work.
5. Correct immediately any obvious problems identified.

Standardize the Process

To continually improve a process, all people involved in its operation must be using the same procedures. Often this is not the case. Employee X may use different procedures than Employee Y. It is important to ensure that all employees are using the best, most effective, most efficient procedures known. The steps involved in standardizing a process are as follows:

1. Identify the currently known best practices and write them down.
2. Test the best practices to determine whether they are, in fact, the best, and improve them if there is room for improvement (these improved practices then become the final best practices that are recorded).

3. Make sure that everyone is using the newly standardized process.
4. Keep records of process performance, update them continually, and use them to identify ways to improve the process even further on a continual basis.

Eliminate Errors in the Process

The strategy of eliminating errors in the process involves identifying errors that are commonly made in the operation of the process and then eliminating them. This strategy helps eliminate steps, procedures, and practices that are being done a certain way simply because that is the way they have always been addressed. Whatever measures can be taken to eliminate such errors are done as a part of this strategy.

Streamline the Process

The strategy of streamlining the process is used to take the slack out of a process. This can be done by reducing inventory, reducing cycle times, and eliminating unnecessary steps. After a process has been streamlined, every step in it has significance, contributes to the desired end, and adds value.

Reduce Sources of Variation

The first step in the strategy of reducing sources of variation is identifying sources of variation. Such sources can often be traced to differences among people, machines, measurement instruments, material, sources of material, operating conditions, and times of day. Differences among people can be attributed to levels of capability, training, education, experience, and motivation. Differences among machines can be attributed to age, design, and maintenance. Regardless of the source of variation, after a source has been identified, this information should be used to reduce the amount of variation to the absolute minimum. For example, if the source of variation is a difference in the levels of training completed by various operators, those who need more training should receive it. If one set of measurement instruments is not as finely calibrated as another, they should be equally calibrated.

Bring the Process under Statistical Control

This strategy of bringing the process under statistical control is explained in detail in Chapter 11. At this point, it is necessary to know only that a control chart is planned, data are collected and charted, special causes are eliminated, and a plan for continual improvement is developed.

Improve the Design of the Process

There are many different ways to design and lay out a process. Most designs can be improved on. The best way to improve the design of a process is through an active program of experimentation. To produce the best results, an experiment must be properly designed, in the following steps:

1. Define the objectives of the experiment. (What factors do you want to improve? What specifically do you want to learn from the experiment?)
2. Decide which factors are going to be measured (cycle time, yield, finish, or something else).
3. Design an experiment that will measure the critical factors and answer the relevant questions.
4. Set up the experiment.

5. Conduct the experiment.
6. Analyze the results.
7. Act on the results.

ADDITIONAL IMPROVEMENT STRATEGIES

In his book *Total Manufacturing Management,* Merli lists 20 strategies for continuous improvement that he calls "The Twenty Organizing Points of Total Manufacturing Management."[17] These strategies are explained in the following paragraphs:[18]

■ *Reduced lead time.*   Raw materials sitting in a storeroom are not adding value to a product. Efficient management of the flow of materials is essential to competitiveness. Lead time can be reduced by evaluating the following factors: order processing time, waiting time prior to production, manufacturing lead time, storage time, and shipping time.

■ *Flow production.*   Traditionally, production has been a stop-and-go or hurry-up-and-wait enterprise. *Flow production* means production that runs smoothly and steadily without interruption. An example illustrates this point. A large manufacturer of metal containers had its shop floor arranged by type of machine (cutting, turning, milling, etc.). All cutting machines were grouped together, all turning machines were grouped together, and all milling machines were grouped together. However, this isn't how the flow of work went. Work flowed from cutting to turning, back to cutting, and on to milling. Arranging machines by type caused a great many interruptions and unnecessary material handling. To improve production efficiency, the machines were rearranged according to work flow. This smoothed out the rough spots and made work flow more smoothly.

■ *Group technology.*   Traditional production lines are straight. With group technology, processes are arranged so that work flows in a U-shaped configuration. This can yield the following benefits: shorter lead times, greater flexibility, less time in material handling, minimum work in progress, flexibility with regard to volume, less floor space used, and less need for direct coordination.

■ *Level production.*   This involves breaking large lots into smaller lots and producing them on a constant basis over a given period of time. For example, rather than producing 60 units per month in one large lot, production might be leveled to produce three units per day (based on 20 work days per month). This strategy can yield the added benefit of eliminating the need to store the materials needed for large lots. This, in turn, makes it easier to implement just-in-time manufacturing.

■ *Synchronized production.*   Synchronized production involves synchronizing the needs of the production line with suppliers of the materials needed on the line. For example, assume that a line produces computers in a variety of different internal configurations. The difference among the configurations is in the capacity of the hard drive installed. Such information as what type of hard drive is needed, in what quantities, at what time, and at what point on the line must be communicated to the hard-drive supplier. The supplier must, in turn, deliver the correct type of hard drive in the correct quantity at the correct time to the correct place on the line. When this happens, synchronized production results.

■ *Overlapped/parallel production.*   This strategy involves dismantling long production lines with large lot capacities and replacing them with production cells that turn out smaller lots. This allows production of different configurations of the same product to be overlapped and/or run parallel.

■ *Flexible schedules.*   Production cells and the ability to overlap production or run it parallel allow for a great deal of flexibility in scheduling. The more options available to production schedulers, the more flexible they can be in developing schedules.

- *Pull control.*   Pull control is a concept applied to eliminate idle time between scheduling points in a production process, the need to maintain oversized inventories to off-set operational imbalances, and the need to plan all target points within a process. With good pull control, work moves through a process uninterrupted by long waiting periods between steps.

- *Visual control.*   Visual control is an important aspect of just-in-time manufacturing. It is an information dissemination system that allows abnormalities in a process to be identified visually as they occur. This, in turn, allows problems to be solved as they occur rather than after the fact.

- *Stockless production.*   Stockless production is an approach to work handling, inventory, lead time planning, process balancing, capacity utilization, and schedule cycling that cuts down on work in progress. With stockless production, it is necessary to eliminate process bottlenecks, balance the process, and have an even work flow that eliminates or at least minimizes work in progress. Stockless production and just-in-time go hand-in-hand.

- *Kidoka.*   *Kidoka* means halting an entire process when a defect is discovered so that it won't cause additional problems further down the line. Kidoka can be accomplished manually, or the line can be programmed to stop automatically, or both.

- *Reduced setup time.*   This strategy consists of any activity that can reduce the amount of time required to break down a process and set it up again for another production run. Such things as quick changeovers of tools and dies are common with this strategy.

- *In-process control.*   Work in process often means work that is sitting idle waiting to be processed. Controlling the amount of idle work involves organizing for a smoother flow, small lot sizes, process flexibility, and rapid breakdown and setup.

- *Quality improvement.*   In addition to improving productivity using the various strategies discussed in this chapter, it is important simultaneously to improve quality. This book is devoted to an approach for continuously improving quality. The important point is that productivity and quality must be improved simultaneously.

- *Total cost cycles.*   This strategy involves basing decisions on the total cost cycle rather than isolated pieces of it. It is not uncommon for decisions to be based on reducing the costs associated with part of a process. True improvements have not been accomplished unless overall costs have been reduced.

- *Cost curves.*   A cost curve is a graphic representation of a time-based process wherein manufacturing costs accumulate relative to billing. Two types of costs are shown on a cost curve: materials and conversion costs. A cost curve shows graphically how much cost accumulates until the customer is billed for the product. It is a tool to help managers economize on the handling of orders.

- *Mushroom concept.*   This strategy is designed to broaden a company's customer base by creating a product that is open to diversification while, at the same time, being sufficiently standardized to minimize production costs. This is done by holding to standard processes as long as possible in the overall production cycle and adding different features only at the end of the process so that a variety of diversified products mushroom out at the end.

- *Suppliers as comakers.*   This strategy amounts to involving suppliers as partners in all phases of product development rather than keeping them out and revealing your activities to them only through the low-bid process. If tested and trusted suppliers know what you are trying to do, they will be better able to maximize their resources in helping you do it.

- *Total industrial engineering.* Total industrial engineering integrates all elements—organizational, technical, and people related—in an effort to achieve continuous improvement. Total industrial engineering focuses its efforts on the improvement of industrial systems rather than people.

■ *Total productive maintenance.*   Total productive maintenance means maintaining all systems and equipment continuously and promptly all of the time. In a rushed work-place, one of the most common occurrences is slacking off on machine and system maintenance. This is unfortunate because a poorly maintained system cannot achieve the quality and productivity to be competitive. Poor maintenance can result in the following problems: shutdowns from unexpected damage, increased setup and adjustment time, unused up time, speeds below the optimum, increased waste from defects, and production losses during startup procedures.

ENDNOTES

1. Joseph M. Juran, *Juran on Leadership for Quality: An Executive Handbook* (New York: Free Press, 1989), 72.
2. Peter R. Scholtes, *The Team Handbook* (Madison, WI: Joiner Associates, 1992), 5-6–5-9.
3. Juran, 35.
4. Juran, 42–46.
5. Juran, 43.
6. Scholtes, 5-9.
7. Scholtes, 5-9.
8. This section is based on Scholtes, 5-10.
9. This section is based on Scholtes, 5-10.
10. This section is based on Scholtes, 5-11.
11. This section is based on Scholtes, 5-11.
12. Giorgio Merli, *Total Manufacturing Management* (Cambridge, MA: Productivity Press, 1990), 143.
13. Scholtes, 5-17.
14. Juran, 47–48.
15. Juran, 48.
16. Scholtes, 5-54–5-67.
17. Merli, 163.
18. Merli, 153–165.

Benchmarking

Benchmarking is becoming an increasingly popular tool among companies trying to become more competitive, striving for world-class performance. The vast majority of them are actively engaged in benchmarking. Benchmarking is a part of the total quality process, and anyone involved in total quality should have a solid understanding of this subject. This chapter is intended to help readers understand what benchmarking is all about, its benefits and its pitfalls. The chapter includes sufficient information to enable any enterprise to make rational decisions concerning benchmarking, including whether or not to do it, and how to go about it.

Benchmarking was brought to our awareness through Robert C. Camp's 1989 landmark book.[1] Since then a number of variations have developed on the benchmarking theme. We have *benchmarking studies* in which there is no contact with an outside firm—information gained is strictly from the public domain. There is no question that this technique can be useful. It is something that the organization should be doing anyway. Sometimes third-party firms specializing in benchmarking studies are contracted for that work. There is considerable doubt that this is really benchmarking, however.

We also have *competitive benchmarking* in which a competitor's operation is studied from a distance *without* the cooperation of the target firm. The aim is to learn something that can help improve process or product quality. Competitive benchmarking uses data that are publicly available, and once again, it is possible to contract this work to specialist third-party firms. This approach doesn't fit our definition of benchmarking.

Also in use are the unstructured plant visits in which the visitor firm has the intention of learning something that will help with its processes or products. This is often called benchmarking but has more aptly been named "industrial tourism." Such visits have some value, but they do not comprise benchmarking.

Many other variations exist, but the form of benchmarking addressed in this book is what has been called *cooperative benchmarking*, or *process benchmarking*, in which key processes are the focus for radical improvement. This involves a cooperative effort by two firms, the benchmarking firm wanting to bring a substandard process up to the world-class level of the partner firm's process.

BENCHMARKING DEFINED

Benchmarking has been around since the early 1980s, but it wasn't until the early 1990s that it became a widely accepted means of improving company performance. In 1985, almost no benchmarking activity existed among the Fortune 500 companies. By 1990, half the Fortune 500 were using this technique. Today companies large and small are finding benchmarking to be an effective component in their total quality effort. If there is a single most likely reason for the slow rise in benchmarking popularity, it is a misunderstanding of the concept—misunderstanding of what benchmarking is, what it is not, and how to do it. It helps to begin with an examination of what benchmarking is not.

Benchmarking Is Not

Cheating	Illegal
Immoral	Industrial espionage
Unethical	

All of these misconceptions about benchmarking assume that one party somehow takes advantage of an unsuspecting competitor by surreptitiously copying the competitor's product or processes. Nothing could be further from the truth. Benchmarking involves two organizations that have previously agreed to share information about processes or operations. The two organizations both anticipate some gain from the exchange of information. Either organization is free to withhold information that is considered proprietary. In addition, the two companies need not be competitors.

> Benchmarking is the process of comparing and measuring an organization's operations or its internal processes against those of a best-in-class performer from inside or outside its industry.

Benchmarking is finding the secrets of success of any given function or process so that a company can learn from the information—and improve on it. It is a process to help a company close the gap with the best-in-class performer without having to reinvent the wheel.

A distinction exists between benchmarking and competitive analysis. *Competitive analysis* involves comparing a competitor's product against yours. It compares the features and pricing of the product. Consumers perform competitive analysis when they compare competitors' products as they try to determine which brand of VCR, television, or automobile to purchase. Benchmarking goes beyond that to comparing how the product is engineered, manufactured, distributed, and supported. Benchmarking is interested not so much in what the product is and what it costs but rather in the underlying processes used to produce, distribute, and support it.

Finally, and most important, benchmarking is a tool to help establish where improvement resources should be allocated. If, for example, it is discovered that three of five processes are nearly as good as the best-in-class performers, but two are significantly off the best-in-class mark, then the most resources should be allocated to these two. The most benefit for the dollars invested will come from changing those processes to conform more nearly to the best-in-class. Relatively little will be gained by drastically changing a process that is already close to the best there is. Key points to remember about benchmarking are as follows:

- Benchmarking is an increasingly popular improvement tool.
- Benchmarking concerns processes and practices.
- Benchmarking is a respected means of identifying processes that require major change.
- Benchmarking is done between consenting companies that may or may not be competitors.
- Benchmarking compares your process or practice with the target company's best-in-class process or practice.
- The goal of benchmarking is to find "secrets of success" and then adapt and improve them for your own application.
- Benchmarking is equally beneficial for both large and small businesses.

RATIONALE FOR BENCHMARKING

The future for companies today seems far different from what it has been since World War II. The first real questions regarding the future and the ability of the United States to sustain our industrial leadership seem to have resulted from the oil crisis of 1974. By then, the United States had lost much of the commercial electronics business to Sony, Hitachi, and Panasonic, but the most important industry in the United States, the automobile, seemed secure. However, when the oil embargo struck, Americans quickly traded their big domestic cars for small, fuel-efficient Japanese models. When the embargo ended, Americans continued buying Japanese cars because at that time they were better than their American counterparts. The Japanese quickly claimed about 30% of the U.S. automobile market (and possibly could have gained much more except for voluntary restraints adopted out of fear that severe trade restrictions would be imposed by Washington). Following these events, North America finally started to wake up to the fact that the world was changing. Whole industries were moving from one part of the world to another, and most of that movement was to Japan. There was good reason to look at Japan to see what they were doing differently that allowed them to accomplish this.

What was learned, of course, was that by following the teachings of Deming, Juran, Ishikawa, Taguchi, Ohno, and other quality pioneers, Japan had developed vastly superior practices and processes. These resulted in superior manufactured goods at competitive prices—everything from motorcycles, to cars, to cameras, to electronics of all kinds, even to shipbuilding. It took several years of looking at Japan to realize fully what had happened. For a long time, Western leaders rationalized Japan's success to low labor costs, the Japanese work ethic versus that of Detroit, lifetime employment, and other factors. Such rationalizations simply clouded the real issue: the superiority of the Japanese practices and processes. Now that industrial leaders worldwide are aware that better practices and processes can enhance competitiveness, it makes good business sense to determine where an organization stands relative to world-class standards and what must be done to perform at that level. That is what benchmarking is designed to do.

Twenty years ago, benchmarking was a case of comparing North American industry with the Japanese. Today, benchmarking is a case of comparing your company with the best in the world. The best in the world for a given comparison may be in Japan, or it may be next door. It may be your direct competition, or it may be in a completely different industry. In addition to companies all over the world emulating the Japanese, customers all over the world are demanding the highest quality in the products they buy. Business as usual will no longer flourish. Organizations must be improving always and forever, or they will be out of business soon and forever.

The rationale for benchmarking is that it makes no sense to stay locked in an isolated laboratory trying to invent a new process that will improve the product, or reduce cost, when that process already exists. If one company has a process that is four times as efficient, the logical thing for other companies to do is to adopt that process. An organization can make incremental improvements to its process through continuous improvement, but it might take years to make a 4× improvement, and by then the competition would probably be at 6× or better. Benchmarking is used to show which processes are candidates for continuous (incremental) improvement and which require major (one-shot) changes. Benchmarking offers the fastest route to significant performance improvement. It can focus an entire organization on the issues that really count.

Some factors that drive companies to benchmark are commitment to total quality, customer focus, product-to-market time, manufacturing cycle time, and financial performance at the bottom line. Every company that has won the Malcolm Baldrige Award endorses benchmarking. Key points to remember about benchmarking as it relates to continuous improvement are as follows:

- Today's competitive world does not allow time for gradual improvement in areas in which a company lags way behind.
- Benchmarking can tell a firm where it stands relative to best-in-class practices and processes, and which processes must be changed.
- Benchmarking provides a best-in-class model to be adopted or even improved on.
- Modern customers are better informed and demand the highest quality and lowest prices. Companies have a choice to either perform with the best or go out of business.
- Benchmarking supports total quality by providing the best means for rapid, significant process/practice improvement.

BENCHMARKING APPROACH AND PROCESS

The benchmarking process is relatively straightforward, but steps must flow in a sequence. A number of variations are possible, but the process should follow this general sequence:

1. Obtain management commitment.
2. Baseline your own processes.
3. Identify your strong and weak processes and document them.
4. Select processes to be benchmarked.
5. Form benchmarking teams.
6. Research the best-in-class.
7. Select candidate best-in-class benchmarking partners.
8. Form agreements with benchmarking partners.
9. Collect data.
10. Analyze data and establish the gap.
11. Plan action to close the gap/surpass.
12. Implement change.
13. Monitor.
14. Update benchmarks; continue the cycle.

These 14 implementation steps are explained in turn in the following sections.

Step 1: Obtain Management Commitment

Benchmarking is not something one approaches casually. It requires a great deal of time from key people, and money must be available for travel to the benchmarking partners' facilities. Both of those require management's approval. You expect to gain information from your benchmarking partner for which they will expect payment in kind, namely, information from you about your processes. This can be authorized only by management. Finally, the object of benchmarking is to discover processes to replace yours or at least to make major changes to them. Such changes cannot be made without management's approval. Without a mandate from top management, there is no point in attempting to benchmark. That is why the requirement for management commitment is at the top of the list. If you cannot secure that commitment, proceed no further.

Step 2: Baseline Your Own Processes

If your company is involved in total quality, chances are good that you have already done some baselining of processes, because before continuous improvement can be used effectively, and certainly before statistical process control can be applied, the processes in question must be understood. That is, the processes must be characterized in terms of capability, their flow diagrams, and other aspects. If this has not been done before, it must be done now. It is critical that you understand your own processes thoroughly before attempting to compare them with someone else's. Most people think they know their processes well, but that is rarely the case if a deliberate process characterization has not been done. It is also important that an organization's processes be completely documented, not just for its own use but for the benefit of everyone associated with the process in any way. (See the discussion of flowcharting in Chapter 9.)

Step 3: Identify and Document Both Strong and Weak Processes

Strong processes will not be benchmarked initially; continuous improvement techniques will be sufficient for them. Weak processes, however, become candidates for radical change through benchmarking, because incremental improvement would not be sufficient to bring them up to the level necessary in the time frame required.

It can be difficult to categorize an organization's processes as weak or strong. A process that creates 50% scrap is an obvious choice for the benchmarking list. On the other hand, a process may be doing what is expected of it and as a result be classified as strong. However, it could be that expectations for that process are not high enough. It is possible that someone else has a process that is much more efficient, but you just don't know about it. Never consider a process to be above benchmarking, no matter how highly it is rated. Concentrate on the weak ones, but keep an open mind about the rest. If research identifies a better process, add it to the list.

Above all, document all processes fully—even the strong ones. Keep in mind that as you are looking at one of your benchmarking partner's processes because it is superior to yours, they will be looking at your strong processes for the same reason. If the processes are not well documented, it will be very difficult to help your partner. It is impossible to compare two processes for benchmarking if both are not fully documented.

Step 4: Select Processes to Be Benchmarked

When you have a good understanding of your own processes and the expectations of them, decide which ones to benchmark. An important point to remember is this: never benchmark a process that you do not wish to change. There is no point in it. Benchmarking is not something you engage in simply to satisfy curiosity. The processes that are put on a benchmark list should be those known to be inferior and that you intend to change. Leave the others for incremental change through continuous improvement—at least for the time being.

Step 5: Form Benchmarking Teams

The teams that will do the actual benchmarking should include people who operate the process, those who have input to the process, and those who take output from it. These people are in the best position to recognize the differences between your process and that of your benchmarking partner. The team must include someone with research capability because it will have to identify a benchmarking partner, and that will require research. Every team should have management representation, not only to keep management informed but also to build the support from management that is necessary for radical change.

Step 6: Research Best-in-Class

It is important that a benchmarking partner be selected on the basis of being best-in-class for the process being benchmarked. In practical terms, it comes down to finding the best-in-class-you-can-find-who-is-willing. Because benchmarking is accomplished by process, best-in-class may be in a completely different industry. For example, say that an organization manufactures copy machines. It might consider potential benchmarking partners who are leaders in the copying industry. But if it is a warehousing process that is to be benchmarked, the company might get better results by looking at catalog companies that have world-class warehousing operations. If the process to be benchmarked is accounts receivable, then perhaps a credit card company would be a good partner.

Processes are shared across many industries, so don't limit research to like industries or you might miss the best opportunities for benchmarking. Remember that best-in-class does not mean best-in-your-industry, but best regardless of industry for the process in question. If team members stay up-to-date with trade journals, they should be able to compile a good list of potential benchmarking partners. Research should cover trade literature, suppliers and customers, Baldrige Award winners, and professional associations. The Internet offers a seemingly endless stream of benchmarking information. Team members will find that the best-in-class processes become well known very fast.

Step 7: Select Candidate Best-in-Class Benchmarking Partners

When the best-in-class have been identified, the team must decide which among them it would prefer to work with. Consideration must be given to location and to whether the best-in-class is a competitor (remember, the team will have to share information with its partner). The best benchmarking partnerships provide some benefit for both parties. If the team can find a way to benefit its potential partner, the linkage between the two companies will be easier to achieve. Even without that, most companies with best-in-class processes are willing, often eager, to share their insights and experience with others, even if they gain nothing in return. Indeed, Baldrige Award winners are expected to share information with other U.S. organizations.

Step 8: Form Agreements with Benchmarking Partners

After the team has selected the candidates, it contacts the potential partner to form an agreement covering benchmarking activities. It can be useful to have an executive contact an executive of the target company, especially if there is an existing relationship or some other common ground. Often the most difficult part of the process is identifying the right person in the potential partnering company. Professional associations can sometimes provide leads to help the team contact someone in the right position with the necessary authority.

After such a contact has been made, the first order of business is to determine the company's willingness to participate. If they are not willing, the team must find another candidate. When a company is willing to participate, an agreement can usually be forged without difficulty. The terms will include visit arrangements to both companies, limits of disclosure, and points of contact. In most cases, these are informal. Even so, care must be exercised not to burden either benchmarking partner with excessive obligations. Make the partnership as unobtrusive as possible.

Step 9: Data Collection

The team has already agreed to discuss a specific process (or processes). Observe, collect, and document everything about the partner's process. In addition to that, try to determine the underlying factors and processes: what is it that makes them successful in this area? For example, do they employ total productive maintenance, continuous improvement, employee involvement, use of statistics, and various other approaches? Optimally, your process operators should talk directly with their operators. It is important to come away with a good understanding of what their process is (flow diagram), its support requirements, timing, and control. The team should also try to gain some understanding of the preceding and succeeding processes, because if you change one, the others may require change as well. If the team knows enough when it leaves the partner's plant to implement its process back home, then it has learned most of what is needed. Anything less than this, and the team has more work to do.

While you are in a partner's plant, try to get a feel for how they operate. Be open-minded and receptive to new ideas that are not directly associated with the process in question. Observing a different plant culture can offer a wealth of ideas worth pursuing.

Step 10: Analyze the Data; Establish the Gap

With the data in hand, the team must analyze it thoroughly in comparison with the data taken from its own process. In most cases, the team will be able to establish the gap (the performance difference between the two processes) numerically—for example, 200 pieces per hour versus 110 pieces, 2% scrap versus 20%, or errors in parts per million rather than parts per thousand.

After the team concludes there is no doubt that the partner's process is superior, other questions arise: Can their process replace ours? What will it cost, and can we afford it? What impact will it have on adjacent processes? Can we support it? Only by answering these questions can the team conclude that implementation is possible.

Step 11: Plan Action to Close the Gap or Surpass

Assume the team concluded that the change to the new process is desirable, affordable, and supportable and that the team wants to adopt the process. In most cases, implementation will require some planning to minimize disruption while the change is being made and while the operators are getting used to the new process. It is very important to approach implementation deliberately and with great care. This is not the time for haste. Consider all conceivable contingencies and plan to avoid them, or at least be prepared for them. Physical implementation may be accompanied by training for the operators, suppliers, or customers. Only after thorough preparation and training should an organization implement the change to the new process.

A second aspect of benchmarking should be kept in mind. The objective is to put in place a process that is best-in-class. If the team merely transplants the partner company's process, it will not achieve the objective, although improvements may occur. To achieve best-in-class, an organization must surpass the performance of the benchmark process. It may not be possible to do this at the outset, but the team's initial planning should provide for the development work necessary to achieve it in a specified period of time (see Figure 13–1).

Step 12: Implement the Change

The easiest step of all may be the actual implementation, assuming that the team's planning has been thorough and that execution adheres to the plan. New equipment may or may not be involved, there may be new people, or more or fewer people—but there

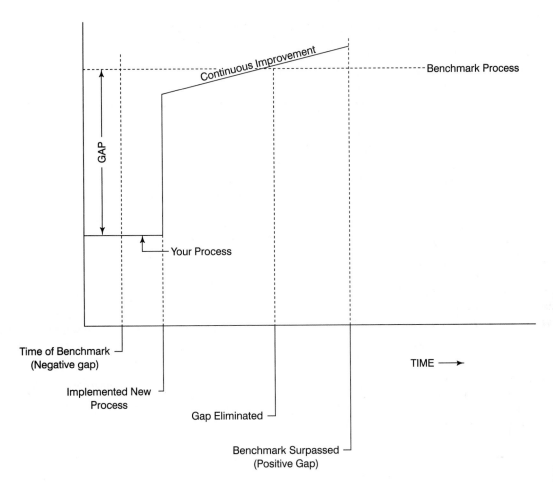

Figure 13–1
Effect of Benchmarking Process Change Followed by Continuous Improvement

will certainly be new procedures that will take time to become routine. Therefore, it should not be a surprise if initial performance does not equal the benchmark. After people get used to the changes and initial problems get worked out, performance should be close to the benchmark. If it is not, an important factor was overlooked, and another visit to the benchmarking partner may be necessary to determine what it is.

Step 13: Monitor

After the process is installed and running, performance should come up to the benchmark quickly. Before long, continuous improvement should enable the organization to surpass the benchmark. None of this is likely to happen without constant attention: monitoring. Never install a new process, get it on-line and performing to expectations, and then forget about it. All processes need constant attention in the form of monitoring. Statistical process control can be an invaluable tool for this purpose, as can other types of charting.

Step 14: Update Benchmarks; Continue the Cycle

As was explained in step 11, the intent of benchmarking is not only to catch up with the best-in-class but to surpass, thereby becoming best-in-class. This is a formidable undertaking, because those with best-in-class processes are probably not resting on their laurels. They too will continue to strive for continually better performance. However, you are now applying new eyes and brains to their process, and fresh ideas may well yield a unique improvement, vaulting your organization ahead of the benchmarking partner. Should that happen, your organization will be sought out as a best-in-class benchmarking partner by others who are trying to bootstrap their performance. Whether that happens or not, whether the benchmark is actually surpassed or not, the important thing is to maintain the goal of achieving best-in-class. Benchmarks must be updated periodically. Stay in touch with the best-in-class. Continue the process. Never be content with a given level of performance.

An important consideration, as you either achieve best-in-class or get close, is that limited resources have to be diverted to those processes that remain lowest in performance relative to their benchmarks. Let continuous improvement take over for the best processes, and concentrate the benchmarking on the ones that remain weak.

Three Phases of Benchmarking

This 14-step sequence represents the three phases of benchmarking: preparation, execution, and postexecution. Figure 13–2 illustrates the benchmarking process/ sequence by phase and indicates action responsibility for each step. Figure 13–2 also makes it clear that the final step (14) causes the cycle to start over again at step 2, confirming the never-ending nature of the benchmarking process for companies that want to achieve or maintain leadership positions. Key points relating to the 14-step sequence of steps for implementing benchmarking are as follows:

- Benchmarking requires top management's commitment, participation, and backing.
- It is necessary that an organization thoroughly understand its own processes before attempting to benchmark.

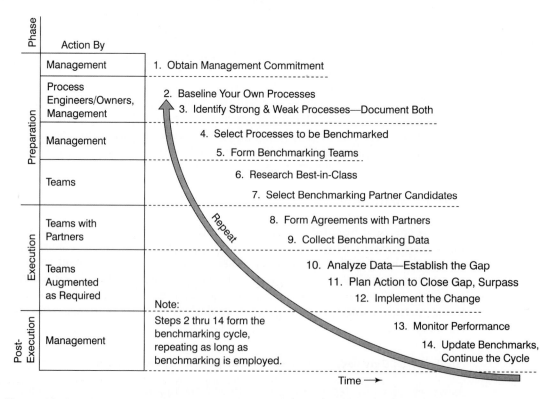

Figure 13–2
The Benchmarking Process/Sequence

- The processes that should be benchmarked are those that need most improvement.
- Benchmarking teams must include process operators.
- Benchmark best-in-class, not best-in-the-industry.
- Do not rush into new processes or major changes without thorough, thoughtful planning.
- Do not be satisfied with a zero gap—aim to surpass.
- Carefully monitor new processes or major process changes.
- Benchmarking is not a one-shot process: continue it forever.

BENCHMARKING VERSUS REENGINEERING

Benchmarking involves partnering with the owner of a best-in-class process so that you might adopt or adapt that process in your operation without having to spend the time and energy to try to design a duplicate of the superior process. Process reengineering requires you to do the latter, on your own. Therefore, in our view, process reengineering should only be considered when it is impossible to use benchmarking. That could happen for a number of reasons, including these:

- No known process available for benchmarking (rare)
- Best-in-class not willing to partner
- Best-in-class inaccessible due to geography or expense

If your subject process is unsatisfactory, and you cannot benchmark for any of these reasons, you may have to resort to reengineering. You should be careful to consider the reasons for the process being unsatisfactory. It may simply be the wrong process for the job, or it may be out of statistical control. Reengineering will not solve either of those problems. Be sure that the process is appropriate, and that it is in control first. If it is still not producing the desired results, suggesting that it is simply not capable, then redesigning it through reengineering is a good approach. One disadvantage with process reengineering is that there is no guarantee that after spending the time and resources, you will have a competitive process. That issue does not exist with benchmarking. With benchmarking you will have observed a competitive process in action.

When we set out to improve our processes, we normally flowchart them to help us understand how the process really works and to give us a visual impression of the steps, people, and functions involved. Improvement typically comes about by changing or eliminating activity in the process that does not add value or consumes too much time or resources, and so forth. There is an alternative way to go about this, and that is to abandon the current process and replace it with a brand-new process that provides the same functionality but better, faster, or cheaper. That is process reengineering.

Here is something to think about. If an organization could achieve the same results by either one of these two routes, which one would stand the best chance for success in the workplace? We believe the former—let's call it the continuous improvement route—would be more readily accepted by the workforce and would be, therefore, more likely to succeed. Usually the people most closely related to the process have a major input to any continuous improvement initiative, and it will not be perceived as something being forced on them by some person or group that doesn't really understand the process anyway. Whether justified or not, that is the way process reengineering has come across to workers. It tends to be radical and sudden, and seldom is consideration given to the human issues. Many times it is seen as a management tool for laying off workers. It does not have to be that way, but that is, we think, the way process reengineering is widely perceived today.

We say this to lead into our final thoughts on process reengineering. If you find process reengineering to be the approach for one of your processes, never let it be a surprise to your employees. In keeping with the philosophy we have promoted throughout this book, it only makes sense to involve the process owners and their internal suppliers and customers, along with other appropriate employees, in your process reengineering project. Take advantage of the collective brainpower and the diverse perspectives; in doing so, their buy-in will be assured.

In summary, if you have a very good process to begin with, use continuous improvement techniques to make it better. On the other hand, if the process is clearly inferior to some used by other firms, try benchmarking. When you cannot achieve the kind of improvement you need from either of those methods, then process reengineering may be required. But no matter which way you go, be sure to get your people involved.

ROLE OF MANAGEMENT IN BENCHMARKING

Management plays a crucial role in the benchmarking process. In fact, without the approval and commitment of top management, benchmarking is not possible. Benchmarking is not something that can occur from the grass roots up without management's direct involvement. Several benchmarking considerations require management's approval before the process can start: commitment to change, funding, personnel, disclosure, and involvement.

Commitment to Change

Benchmarking is a serious undertaking for both benchmarking partners. Unless a firm commitment to change exists—unless the organization fully intends to radically improve its processes to come up to best-in-class standards—benchmarking should not be considered. Unfortunately, too many companies jump into benchmarking without that commitment, with the result that money and personnel are wasted by both parties. In addition, the hopes and expectations of employees are raised, only to be disappointed when nothing comes of it. To obtain any real benefit from benchmarking, an organization must resolve that when a best-in-class process is found, it will do what is necessary to incorporate it as a replacement (or radical improvement) model for its inferior process. That, after all, is what benchmarking is about.

Funding

Only management can authorize the expenditure of funds for benchmarking. These funds will support travel for teams visiting the organizations with best-in-class processes. Teams are usually composed of five to eight people. Visits may last from two days to two weeks. Travel destinations are inflexible, dictated by the location of the best-in-class firms. Clearly, travel expenses can be costly. Management must make the funds available if benchmarking is to be carried out.

Human Resources

In similar fashion to funding, management must make the necessary human resources available for the benchmarking tasks. Although the costs for the human resources are usually far higher than for travel, the availability of personnel is seldom an issue except for the target company.

Disclosure

It may not be immediately obvious, but both companies—the benchmarker and the target—disclose information about their processes and practices. Management may be understandably hesitant to disclose such information to competitors, but what about the case of the noncompetitor benchmarking partner? Even there, management may be reluctant, because there can be no ironclad guarantee that information divugled to a noncompetitor will not find its way to the competition. The other side of the coin is that few processes or practices remain secret very long anyway. But if the organization has some unique process that gives it a competitive advantage, the process should be treated as proprietary and not be subjected to benchmarking. In any event, only management can make the decision to disclose information.

Involvement

Management must be actively and visibly involved in every aspect of the benchmarking process. Management should be involved in determining which processes are to be benchmarked and who the benchmarking partner candidates are. Management is in a unique position to establish the communication channels between the companies, because top managers tend to affiliate through professional organizations. Dialogue among top-level managers should be encouraged.

It is important for management to stay abreast of benchmarking events and to make certain that the effort supports the objectives and vision of the company. Management's ability to do this is greatly enhanced when it is directly involved. In addition, subordinates will recognize the importance placed on benchmarking by the degree to which management is visible in the process. With management active, all levels will be more productive in their benchmarking activities.

PREREQUISITES OF BENCHMARKING

Before getting involved in benchmarking, an organization should check the prerequisites—those philosophical and attitudinal mind-sets, the skills, and the necessary preliminary tasks that must precede any benchmarking efforts.

Will and Commitment

Without the will and commitment to benchmark, an organization cannot proceed. Don't waste time or the time of a benchmarking partner in the absence of a commitment and a will to benchmark on the part of the company's top management.

Vision/Strategic Objective Link

Benchmarking requires a strong focus, or it can go off in numerous different directions as benchmarkers get carried away in their enthusiasm. Before it is started, benchmarking objectives must be linked to the company's vision and strategic objectives, providing specific direction and focus for the effort. Failure to do this will almost certainly result in wasted resources and frustration.

Goal to Become Best—Not Simply Improved

Nothing is wrong with incremental improvement—unless current performance is far below world-class. However, if an organization is not at the world-class level, incremental improvement may only ensure that it remains inferior to the best-in-class forever. Benchmarking requires that the goal be to leap to the head of the field in one radical change, not just to be a few percentage points better than last year.

Openness to New Ideas

If a company is imbued with the not-invented-here syndrome, it will have a problem with benchmarking. The chief symptom of that affliction is a shortsighted mind-set that is characterized by a reluctance to consider other ways of doing things. Although few will admit it, many people are reluctant to consider ideas or approaches that are not their own. Organizations can be like individuals in this regard. Because the essence of benchmarking is capitalizing on the work and ideas of others, a company must be open to new ideas for benchmarking to provide any value. The benchmarking process may help bring about more receptivity to new ideas by demonstrating that they really work.

Understanding of Existing Processes, Products, Services, and Customer Needs

It is mandatory that an organization thoroughly understand its own processes, products, services, practices, and the needs of its customers so that it can determine what needs to be benchmarked. In addition, it is necessary to have a solid understanding of your process in order to make meaningful measurements against that of the partner.

Processes Documented

It is not enough to understand the processes; they must be completely documented, for three reasons:

- All people associated with the process should have a common understanding of it, and that can come only from documentation.
- A documented starting point is needed against which to measure performance improvement after benchmarking changes have been implemented.
- The organization will be dealing with people (the partners) who are not familiar with its processes. With an understanding of where the benchmarking organization is, the partner will be better able to help.

Process Analysis Skills

To achieve an understanding of your own processes, products, and services and to document those processes, you must have people with the skills to characterize and document processes. These same people will be needed to analyze the benchmarking partners' processes and to help adapt those processes to the organization's needs. Ideally, they should be employees, but it is possible to use consultants in this role.

Research, Communication, and Team-Building Skills

Additional skills required include research, communication, and team building. Research is required to identify the best-in-class process owners. Communication and team building are required to carry out the benchmarking both on an internal basis and with the partners.

OBSTACLES TO SUCCESSFUL BENCHMARKING

Like most human endeavors, benchmarking can fail. Failure in any activity usually means that the participant failed to prepare adequately for the venture—failed to learn enough about the requirements, the rules, the pitfalls. So it can be with benchmarking. In this section, some of the common obstacles to successful benchmarking as drawn from the experiences of dozens of companies are explained.

Internal Focus

For benchmarking to produce the desired results, you have to know that someone out there has a far better process. If a company is internally focused (as many are), it may not even be aware that its process is 80% less efficient than the best-in-class. An internal focus limits vision. Is someone better? Who is it? Such organizations don't even ask the question. This is complacency—and it can destroy an organization.

Benchmarking Objective Too Broad

An overly broad benchmarking objective such as "Improve the bottom-line performance" can guarantee failure. This may well be the reason for benchmarking, but the team will need something more specific and oriented not to the *what* but to the *how*. A team could struggle with the bottom line forever without knowing with certainty that it achieved success or failure. The team needs a narrower target like "Refine or replace the invoicing process to reduce errors by 50%." That gives them something they can go after.

Unrealistic Timetables

Benchmarking is an involved process that cannot be compressed into a few weeks. Consider 4 to 6 months the shortest schedule for an experienced team, with 6 to 8 months the norm. Trying to do it in less time than that will force the team to cut corners, which can lead to failure. If you want to take advantage of benchmarking, be patient. On the other hand, any project that goes on for more than a year should be checked. The team is probably floundering.

Poor Team Composition

When a process is benchmarked, those who own the process, the people who use it day in and day out, must be involved. These people may be production line operators or clerks. Management may be reluctant to take up valuable team slots with these personnel when the positions could otherwise be occupied by engineers or supervisors. Engineers and supervisors should certainly be involved, but not to the exclusion of process owners. The process owners are the ones who know the most about how the process really operates, and they will be the ones who can most readily detect the often subtle differences between your process and that of the benchmarking partner. Teams should usually be six to eight people, so be sure the first members assigned are the operators. There will still be room for engineers and supervisors.

Settling for "OK-in-Class"

Too often organizations choose benchmarking partners who are not best-in-class, for one of three reasons:

- The best-in-class is not interested in participating.
- Research identified the wrong partner.
- The benchmarking company got lazy and picked a handy partner.

Organizations get involved in benchmarking when they decide that one or more of their processes is much inferior to the best-in-class. The intention is to examine that best-in-class process and adapt it to local needs, quickly bringing your organization up to world-class standards in that process area. It makes no sense to link with a partner whose process is just good. It may be better than yours, but if adopted, it still leaves your organization far below best-in-class. For the same amount of effort, an organization could have made it to the top. Organizations should identify the best and go for it. Only if the absolute best will not participate can taking second-best be justified. Second-best should be used only if it is significantly superior to the process in question.

Improper Emphasis

A frequent cause of failure in benchmarking is that teams get bogged down in collecting endless data and put too much emphasis on the numbers. Both data collection and the actual numbers are important, of course, but the most important issue is the process itself. Take enough data to understand your partner's process on paper, and analyze the numbers sufficiently to be certain that your results can be significantly improved by implementing the new process. Unless the team has been deeply involved in the process, the practical knowledge to successfully adapt and implement it back home may be lacking. Keep the emphasis on the process, with data and numbers supporting that emphasis.

Insensitivity to Partners

Nothing will break up a benchmarking partnership quicker than insensitivity. Remember that a partner is doing your organization a favor by giving access to its process. You are taking valuable time from the partner's key people, and at best you are disrupting the routine of daily business. If you fail to observe protocol and common courtesy in all transactions, your organization runs the risk of being cut off.

Limited Top Management Support

This issue keeps coming up because it is so critical to success at all stages of the benchmarking activity. Unwavering support from the top is required to get benchmarking started, to carry it through the preparation phase, and finally to secure the promised gains.

BENCHMARKING RESOURCES

A number of sources of information can help organizations with their benchmarking efforts. They cover the spectrum from nonprofit associations to cooperative affiliations to for-profit organizations that sell information. In addition, of course, there are consulting firms with expertise and databases covering all aspects of benchmarking.

One of the most promising ventures is the American Productivity and Quality Center (APQC) Benchmarking Clearinghouse (123 N. Post Oak Lane, Houston, TX 77024; phone, [713] 685-4657; fax, [713] 681-5321). The APQC Benchmarking Clearinghouse has been set up to assist companies, nonprofit organizations, and government in the process of benchmarking. It works with an affiliation of organizations to collect and disseminate best practices through databases, case studies, publications, seminars, conferences, videos, and other media.

A wide range of benchmarking information is available on the Internet. Just ask your search engine to find "benchmarking" or "process benchmarking," and you will probably be rewarded with more information than you can use. It ranges from articles on the subject to promotions for books and consultants. Colleges list the contents of their libraries that are related to benchmarking. We would suggest a word of caution, however. Anyone can put anything on the Web without verification, so it is always a good idea to approach material from unfamiliar sources with a degree of skepticism. In spite of this, we consider the Web to be a valuable benchmarking resources center.

Excellent sources of information for benchmarking are trade and professional organizations. They can often direct organizations to best-in-class practices, provide contacts, and offer valuable advice. Baldrige Award winners are committed to share information with other U.S. companies, and they hold periodic seminars for this purpose.

The trade literature publishes a wealth of relevant information, including lists of companies with best-in-class processes and practices. *Industry Week* is one example of an excellent source of benchmarking information. Companies such as Dun and Bradstreet and Lexis-Nexis maintain databases of potential benchmarking partners and share them for a fee.

Consultants and universities that are engaged in benchmarking can help organizations get started by providing initial training, offering advice and guidance, and directing organizations to benchmarking partner candidates.

A word of caution is in order at this point. Be sure that any information obtained is current. The very nature of benchmarking makes yesterday's data obsolete. To achieve maximum benefit, organizations must be sure that they are operating on current information.

SELECTION OF PROCESSES/FUNCTIONS TO BENCHMARK

Selection of processes or functions to benchmark would seem to be a straightforward decision but is in fact one that gives many would-be benchmarkers a great deal of trouble. If you keep in mind that the purpose of benchmarking is to make a radical improvement in the performance of a process—more improvement than could be made quickly through continuous improvement techniques—then it follows that most concern should be focused on the weakest processes and the functions that operate them.

The strongest processes are sometimes benchmarked as a means of obtaining a report card against the best-in-class. This is a waste of time and effort, to say nothing of money, on two counts. First, the organization is proud of this process and has no intention of replacing it or radically modifying it. What good does it do to determine that the process is within 10% of best-in-class? It may be intellectually gratifying, but the process will be no better for the effort. Second, the processes that are the weakest are the ones that are most detrimental to competitiveness, not those that are in the 90th percentile. Moreover, the weakest offer the most room for dramatic improvement, perhaps many times over. This is where the benchmarking effort should be focused. The reason companies get this wrong is that they are more inclined to talk about what they do right than what they do wrong. When attempting to benchmark, it is a good idea to leave vanity and pride out of the process.

ACTING ON THE BENCHMARK DATA

At the conclusion of the benchmarking project with your partner, data analysis will have produced both quantitative and qualitative information. The quantitative information is effectively the stake driven into the ground as the point from which future progress is measured. It is also used as the basis for improvement objectives. Qualitative information covers such things as personnel policies, training, management styles and hierarchy, total quality maturity, and so on. This information provides insights on how the benchmarking partner got to be best-in-class.

The quantitative data are clearly the information sought and are always used. However, there may be more value in the qualitative information. It describes the atmosphere and environment in which best-in-class can be developed and sustained. Do not ignore it. Take it very seriously. Study it, discuss it in staff meetings, and explore the possibilities of introducing these changes into your culture.

In terms of the process that has been benchmarked, if the partner's process is significantly superior to your own—and we must assume that it is or it would not have been selected in the first place—you have to do something about implementing it. Perhaps you can modify your own process with some ideas picked up from benchmarking, or, more likely, you can adopt your partner's process, implementing it to replace yours. But whatever is indicated by the particular local situation, take decisive action and get it done.

PERPETUAL BENCHMARKING

If you have been through a series of benchmarking activities and have implemented changes that have significantly improved processes, your organization may develop a tendency to leave benchmarking. After all, there are other things that need attention and resources. But this can be a costly mistake. At this point, the organization not only has much-improved processes, but it has developed some valuable benchmarking experience. Keep in mind that best-in-class continues to be a dynamic and ever-changing mosaic. Processes are constantly being improved and altered. In a relatively short time, an organization can fall behind again. To prevent that from happening, the organization must take advantage of hard-won benchmarking experience and keep the effort moving. This means staying up-to-date with best-in-class through all the means at your disposal, staying current with your own processes as they are continually improved, and benchmarking the weaker processes. This is a never-ending process.

ENDNOTES

1. Robert C. Camp, *Benchmarking: The Search for Industry Best Practices That Lead to Superior Performance* (Milwaukee: Quality Press/Quality Resources, 1989).

CHAPTER FOURTEEN

Just-in-Time (JIT)

JUST-IN-TIME DEFINED

Just-in-time is the name given the Toyota Production System developed by Taiichi Ohno. As is so often the case, we find that the same product is being repackaged under other names. JIT is sometimes referred to as *lean production manufacturing*. The term *focused factory* is sometimes applied to JIT production cells. If you encounter a production system called *demand flow*, or *demand flow technology*, it is JIT with a new label. None of these are bad names, and in fact some may be more accurate. But in this book, and most others, the generic name for pull-system manufacturing, *just-in-time*, is preferred.

When people who should know are asked to define just-in-time (JIT), the typical response is that JIT "is getting your materials delivered just when you need them." Probing a little deeper may elicit a response that suggests JIT manufacturers let their suppliers keep their materials inventory until they need it. The first statement demonstrates an inadequate understanding of JIT, and the second is simply wrong. Even so, many companies under the auspices of JIT have indeed pushed their warehousing back to the suppliers for a net gain of zero. If these are not the right answers to the question "What is JIT?" then what is it? Although not exactly what was originally intended, just-in-time manufacturing has become a management philosophy that seeks to eliminate all forms of waste in manufacturing processes and their support activities. JIT permits the production of only what is needed, when it is needed, and only in the quantity needed. This has to apply not only to the just-in-time manufacturer but also to its suppliers if the system is to eliminate all possible waste. Those companies that have required their suppliers to do their warehousing clearly have not gotten the point. The supplier should not produce the material until the JIT manufacturer needs it. In that mode there is no warehousing and therefore no wasted resources for buildings, maintenance, people to care for the material, spoilage, obsolescence, or other related problems.

JIT is not so much related to supplier activities, although they are important, but more to events on the manufacturing floor. For example, assume that a company manufactures motion sensors. There are five discrete processes involved, each attended to by one worker, as illustrated in Figure 14–1. The traditional production process places a big supply of input materials in the warehouse, doling it out to the production line at the rate of so many pieces per unit time. The electronic assembly and mechanical assembly processes convert their respective input materials into input materials for the electronic module assembly process. The electronic module assembly and the frame fabrication processes then convert their input materials into input materials for the final assembly process, which in turn converts them into completed motion sensors. Each of the five work areas produces at the rate necessary to meet a quota, or to consume all the input material. The completed sensors are sent to the warehouse for storage until someone buys them.

Each preceding stage pushes its output into the succeeding stage. It is difficult to balance a line to the point that the succeeding stages need exactly what is produced by the preceding stages, so it is common to take the output of the preceding stages off the floor and store it in so-called staging areas. Staging areas are nothing more than mini-warehouses.

Just-in-time approaches the manufacturing process from the opposite end of the line. Rather than push materials into the processes, storing them whenever they cannot be accommodated, JIT controls the line from the output end. Indeed, it can be said that the customer controls the line, because nothing is built until there is an order for it. After the order is received for a product, the final assembly process is turned on to put together the required number. The assembler pulls the required input materials from the electronic module and frame fabrication processes—only enough to make the required number. Similarly, the electronic module assembly and frame fabrication processes pull input materials from their preceding processes, and so on back up the line. At the top of the line, input materials are pulled from suppliers in the exact quantity needed, and no more.

Following the JIT procedure, no step in the production process ever overproduces or produces before a demand is made. Therefore, there is no need for a staging area or the people required to move materials into it and out of it, account for it, and so on. No money is tied up in inventory of raw materials, partially built goods (known as work-in-process or WIP), or finished goods. If there are no stored materials, there is no spoilage or obsolescence. The elimination of these wastes alone makes JIT the most powerful manufacturing concept to come along since Henry Ford's moving assembly line of 1913. JIT contributes to the elimination of many more forms of waste, which is discussed later in this chapter.

So the definition of JIT as used in this book is as follows:

> Just-in-time (JIT) is producing only what is needed, when it is needed, and in the quantity that is needed.

RATIONALE FOR JIT

Mass production manufacturers set their production schedules based on a forecast of future needs, which in turn is based on historical data and trend analysis (see Figure 14–2). The great weakness of this system is that no one can predict the future with sufficient certainty, even with a complete and perfect understanding of the past and a good sense of current trends in the marketplace. One does not have to search long to find examples of failed attempts to correctly project the marketability of products. The Edsel is one of many automobiles that were released with great fanfare to a disinterested public. A new formula for Coca-Cola introduced in the late 1980s is another example of market predictions gone awry. IBM has case after case of personal computers, such as the unlamented IBM PC Jr. (which failed in the marketplace in spite of the best market research IBM could muster). These failures demonstrate the difficulty in trying to determine before-hand what will sell and in what quantity.

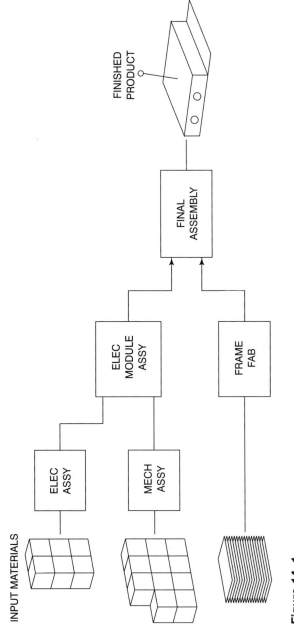

Figure 14–1
The Traditional Production Process

233

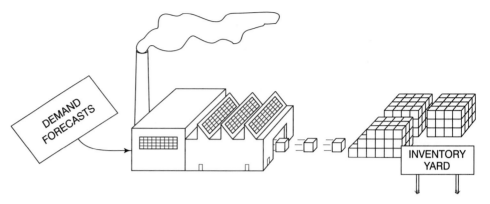

Figure 14–2
Factory Producing to Forecast Demand (Mass Production)

Even products that are successful in the market have limits as to the quantities that will be absorbed by buyers. When production is based on predictions of the future, risk of loss from overproduction is far greater than when production is based on actual demand. The previous section defined JIT as producing what is needed, when it is needed, and only in the quantity that is needed (see Figure 14–3). The result of JIT is that no goods are produced without demand. This, in turn, means no goods are produced that cannot be sold at a price that supports the viability of the company.

So far we have viewed JIT from the point of view of the manufacturer and the ultimate purchaser of the product—the producer and the customer. But if we look at the complete production process, we will find that it contains many producers and customers—internal customers (see Figure 14–4). Each preceding process in the overall system is a producer, or supplier, and each succeeding process is a customer (see Chapter 4). JIT fits here as well as or better than with the manufacturer-and-purchaser model. No process in the system produces its output product until it is signaled to do so by the succeeding process. This can eliminate waste on a grand scale. It is the elimination of waste that justifies JIT in any kind of manufacturing operation. Elimination of waste is translated to improved quality and lower cost. Improved quality and lower costs translate to becoming more competitive. Although improved competitiveness does not assure survival (the competition may still be ahead of you), being noncompetitive surely guarantees disaster.

Taiichi Ohno, the creator of the just-in-time system, saw that the mass production system produced waste at every step of the way. He identified seven wastes:[1]

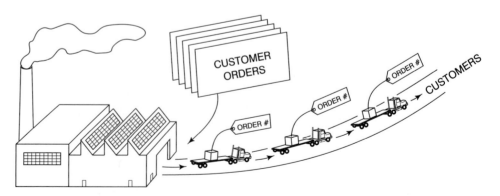

Figure 14–3
Factory Producing to Orders (JIT)

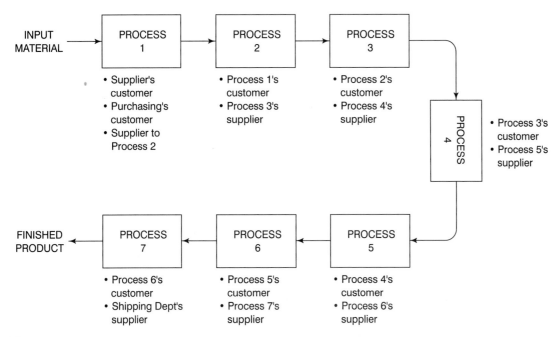

Figure 14–4
Internal Supplier-Customer Relationships

1. Overproducing
2. Waiting (time)
3. Transporting
4. Processing itself
5. Having unnecessary stock on hand
6. Using unnecessary motion
7. Producing defective goods

The elimination of these wastes is at the heart of the rationale for just-in-time: eliminate these wastes, and you will produce better products at lower cost. If the competition gets there first, your rationale for JIT is survival.

DEVELOPMENT OF JUST-IN-TIME

In the preceding section, Taiichi Ohno was identified as the creator of the just-in-time system, and it is true that he was responsible for developing the system as it is now known. However, other names should be added, at least to the extent to which they contributed by inspiration. The first is Henry Ford, creator of mass production. Because of Ford's great appreciation of the expense of waste, Ohno believes that if he were alive today he would be using a system much like Toyota's. In his 1926 book *Today and Tomorrow*, Henry Ford talked about the waste of inventory in raw materials, work in process, and finished goods in the pipeline to market—and about the efforts taken to reduce the investment in this waste. Between 1921 and 1926 Ford output doubled, but investment in inventory of raw materials, semifinished, and finished goods actually declined. Based on 1921 performance, Ford should have had $170 million tied up in this inventory but in fact had less than $50 million. Ford also recognized the wastes arising from transportation, waiting (time), and inefficiency on the factory floor. He believed in planning ahead to eliminate the waste before it happened. This is very contemporary thinking, and Ohno may be correct that Henry Ford, had he been living in the past 30 years, might well have developed a Toyota-like system. When Ohno wrote his book on the Toyota Production System, it was entitled *Just-in-Time for Today and*

Tomorrow. It is not known if this was a tribute to Henry Ford's book, but it is at least an interesting coincidence.

Ford was a great influence on the Toyoda family—Sakichi, Kiichiro, and Eiji. Sakichi Toyoda, a designer of looms and founder of Toyota, is credited with the concept of *autonomation*, or automation with a human touch. His automatic loom could determine whether a thread was broken or missing, shutting itself down instead of making defective product.[2] Autonomation is one of the two pillars of the Toyota Production System, the other being just-in-time. Kiichiro Toyoda, Toyota's founding chairman, planted the seeds of the Toyota Production System prior to World War II with his planning for the introduction of the assembly line at Toyota's Kariya plant. He wrote a booklet about how production was to work, and it contains the words just-in-time. His original meaning in English was "just-on-time," intending that things be done exactly on schedule, with no surplus produced. World War II halted further work on the system, and after the war it was Taiichi Ohno who revived and developed it into the present-day Toyota Production System.[3]

Eiji Toyoda, Toyota's chairman and Taiichi Ohno's boss for 35 years, is credited with the JIT philosophy: "In broad industries, such as automobile manufacturing, it is best to have the various parts arrive alongside the assembly line just-in-time."[4] Eiji Toyoda's greatest contribution may have been his support for Ohno's trial-and-error approach, shielding him from the inevitable controversy of his endeavors. Ohno claims that Eiji never told him to back off or slow down. He absorbed the heat and let Ohno press on unimpeded.[5]

Taiichi Ohno's motivation, like the Toyodas', was to eliminate all forms of waste from the production process. He was well schooled in the Ford mass production system and observed that the system itself created waste in huge proportions. If one was determined to violate the seven wastes, a mass production line would do it. Mass production is prone to *overproducing*, having people or materials *waiting, moving* work in process back and forth across the plant, retaining inefficient *processes*, maintaining costly inventories of *stock on hand*, requiring non-value-added *movement* because lines were set up to accommodate product, not workers, and producing *defective* goods because the line must continue to move. The words in italics represent the seven wastes.

Ohno believed that a production system based on just-in-time could eliminate the wastes. To appreciate fully what is involved here, one must understand that the mass production system as defined by Henry Ford was not irrational. Ford's objective was to produce huge quantities of the same product using an assembly line technology that required little expertise of its workers. The result was a reliable, cheap car that millions of buyers could afford. In that, he and others who used his mass production technology were eminently successful. But mass production is inflexible and wasteful—inflexible because it is driven by the great stamping presses and other machines that do not easily accommodate a variety of products, and wasteful because the underlying philosophy of mass production is that the line must crank out products that spring from market forecasts in a never-ending high-volume stream. To support that high-volume stream, there must be stockpiles of the materials that go into the product, because the lack of a single part can shut down the mass production line. Machines must be capable of high output and are so costly they cannot sit idle without creating trauma in the accounting department. Therefore, even when fenders are not needed, the machines must continue to stamp them out. The overproduction will be warehoused until it is needed—perhaps when the press breaks down. So it is with all the parts and subassemblies that make up the complete product. They are stored in large quantities, just in case something goes wrong in their production or transportation cycle, when they might be needed to keep the final assembly line moving—fenders for a rainy day, so to speak.

This is the norm with mass production. The problem with this system is that the building space in which these parts and materials are warehoused is expensive. It requires a small army of people to care for the stored materials/parts, and these people add not a whit to the ultimate value of the product. Spoilage occurs by loss, damage, or obsolescence for stored parts—all waste. Part waste of inventory, part waste of over-production.

Mass production advocates emphasize the need to keep the line moving and that the only way to do this is to have lots of parts available for any contingency that might arise. This is the fallacy of just-in-time according to mass production advocates. JIT, with no buffer stock of parts, is too precarious. One missing part or a single failure of a machine (because there are no stores of parts) causes the JIT line to stop. It was this very idea that represented the power of JIT to Ohno. It meant that there could be no work-arounds for problems that did develop, only solutions to the problems. It served to focus everyone concerned with the production process on anticipating problems before they happened, and developing and implementing solutions so that they would not cause mischief later on.[6] The fact is that as long as the factory has the security buffer of a warehouse full of parts that might be needed, problems that interrupt the flow of parts to the line do not get solved because they are hidden by the buffer stock. When that buffer is eliminated, the same problems become immediately visible, take on a new urgency, and solutions emerge—solutions that not only fix the problem for this time, but for the future as well. Ohno was absolutely correct. JIT's perceived weakness is one of its great strengths.

Mass production is a *push system* (see Figure 14–5). The marketing forecast tells the factory what to produce and in what quantity; raw materials and parts are purchased, stored, forced into the front end of the production process, and subsequently pushed through each succeeding step of the process, until finally the completed product arrives at the shipping dock. It is hoped that by then there are orders for these goods, or they will have to be either stored or pushed (forced) into the dealers' hands, a widespread practice in the automobile business. The whole procedure, from imperfect forecast of marketability to the warehouse or the dealer is one of pushing.

What if the market will only take half of the predicted amount or wants none? What if the final assembly process can accommodate only two-thirds of the preceding processes' output? These situations present big problems in terms of cost and waste, and they are common.

Just-in-time, on the other hand, is a *pull system* (see Figure 14–6; the term *kanban* in the figure will be clarified soon). The production schedule does not originate in a market forecast, although a great deal of market research is done to determine what customers want. The production demand comes from the customer. Moreover, the demand is made on the final assembly process by pulling finished products out of the factory. The operators of that process in turn place their pull demands on the preceding process, and that cycle is repeated until finally the pull demand reaches back to the material and parts suppliers. Each process and each supplier is allowed to furnish only the quantity of its output needed by the succeeding process.

Figures 14–5 and 14–6 also show a difference in the relationship between the customer and the factory. In the mass production system, no real relationship exists at all. The market forecasters take the place of the customers and place demands on the factory months in advance of production. In the JIT system, however, the customer's demand is felt throughout the system, all the way to the factory's suppliers, and even beyond that. The JIT system is simpler, eliminating whole functions such as material control, production control, warehousing/stocking, and so on.

The simplicity of JIT production is most evident on the factory floor. In a mass production plant, or even a conventional job shop (low-volume, high-variety shops), it is almost never possible to tell from the factory floor how things are going relative to schedules. Parts of any product may be in any number of disparate locations in the plant at any given time—in the machine shop, welding shop, on the line, or in storage. Computers keep track of it all, but even then it is difficult to track a given product through

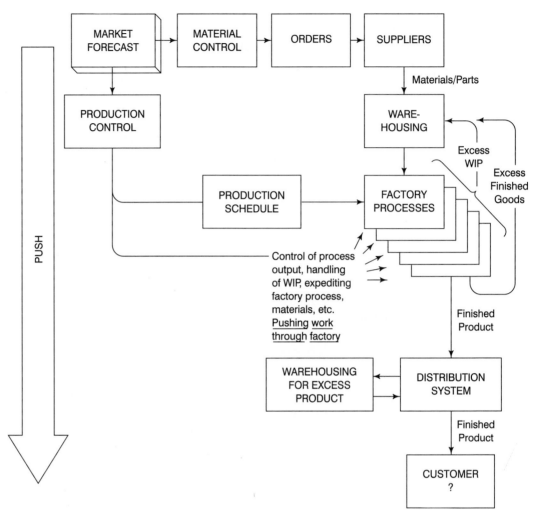

Figure 14–5
Mass Production Push System

the plant or to track its status at a given point in time. On the other hand, JIT, being a very visual process, makes tracking easy—even without computers. Parts have no place to hide in a JIT factory. The only work in process is that for which the process has a kanban.

The simplicity of today's JIT belies the difficulty Ohno encountered in developing the system. Because production must stop for a missing part, a process problem, or a broken machine, methods had to be developed to prevent these occurrences. These preventive strategies are explained in the following sections.

Machine Problems

There are two basic concerns about machines:

1. Is it running and turning out product?
2. If it is running, is the quality of its output product acceptable?

In a mass production environment, question 1 matters most. The tendency is to let the machine run as long as there is product, good or bad, coming out of it. Defec-

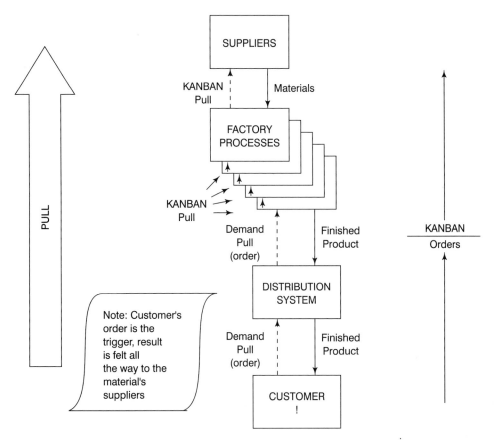

Figure 14–6
Just-in-Time Demand Pull System

tive parts will cause problems farther down the line, but the consequences of shutting the machine down to fix it are seen as an even bigger problem. The JIT factory is more concerned about the second question, because allowing a machine to produce defective parts is to permit the production of waste, and that, above all, is forbidden.

Common sense dictates that machinery should always be maintained properly, but that can be very difficult in a mass production plant. Unfortunately, in many North American factories, machines tend to be ignored until they break down, in keeping with the grammatically incorrect but telling expression "If it ain't broke, don't fix it." Toyota eliminated the machine problem through a systematic preventive maintenance process that keeps all machinery in top shape, modifying it for better reliability or performance, and even predicting when parts should be replaced or adjustments made to maintain the highest-quality output. This has come to be known as *total productive maintenance* or *total preventive maintenance* (TPM). It is finding widespread acceptance in the forward-looking companies. TPM, by keeping the machines available for use when they are needed, eliminates a great many line stoppages. We will discuss TPM in more detail later in the chapter.

Process Problems

Process problems can be eliminated when people thoroughly understand the processes, optimize them, and use statistical methods (i.e., SPC) to keep them under control. In addition, the processes are continually improved, most often through the efforts of the same people who work with them every day. Time is allocated for these kinds of efforts in all JIT factories.

The most difficult conceptual problem with JIT is the precise control of production and material/parts flow through the complete production process. For that, Ohno developed the use of the *kanban* to signal the pulls through the system. Mass production demonstrated that one should not start the control at the beginning of the process. Too many things can go wrong at the bow wave of the flow. Ohno decided that the control had to start at the output end of the factory. From this concept, he introduced kanban. *Kanban* is a Japanese word meaning "card." Ohno used kanbans to trigger activity and the flow of materials/parts from one process to another. When a succeeding process uses the output of the preceding process, it issues a kanban to the preceding process to produce another.

Although Ohno describes the kanbans as slips of paper in a vinyl pouch—close enough for "card"—kanbans have evolved to a number of forms. A square painted or taped on a workstation may be a very effective kanban. For example, a process produces a subassembly and places it on the marked area of the succeeding process workstation. When the succeeding process uses the subassembly, the marked area, the kanban square, becomes empty and signals the preceding process to make another subassembly and fill the square. The same is done with totable bins. When the parts from a bin have been used, the empty bin is sent back to the preceding process as a signal for more production. Both of these kanban devices work when the part or subassembly in question is the only possible output of the preceding process. Should there be a variety of part or subassembly models, however, the kanban square alone will not provide sufficient information, and the bin with a descriptive card or the kanban card must be used. (More information about kanban is provided later in this chapter.)

Lot Size

A final issue for JIT production to overcome had to do with lot size. Mass production is keyed to the largest possible lot sizes: set up the machines and parts streams to make as many as possible of the same item, like Henry Ford's identical black Model T's, before changing to another model or product. So-called economic lot size is still being taught in many universities. Just-in-time seeks to build in the smallest possible lots. The modern consumer demands variety. No auto company could survive today with a single car model, with each unit the same in all respects including equipment and color. JIT accommodates variety by being flexible. That is, the factory is set up so that changes can be rapidly implemented, and with small cost.

Traditionally, it has been a major problem to change models on a production line because breakdown and setup of the machines that have to be changed takes a lot of time. Hours and days and even longer for new setups are not uncommon. Ohno saw that the inherent inflexibility of the mass production line was in the setup time for the machines. Too much setup time meant that a manufacturer had to have a second line— or even a new factory—for the other model, or the customers' demand for the second model was simply ignored until the run on the current model was finished. By attacking the problem head-on, Toyota was able to reduce setup times to the point where they were no longer significant. Other companies, using the Toyota approach, found that they could quickly reduce setup times by 90% and even more with some effort.

Omark Industries was one of the first American companies to study the Toyota Production System. Using Toyota's techniques, they reduced the setup time for a large press from 8 hours to 1 minute and 4 seconds.[7] After setup time became a nonfactor, it was possible to manufacture in small lots—even lots of one—thereby permitting the intermixing of models on the same line. This ability is crucial if the factory is to respond to customer demand in a pull system. This meant that customer responsiveness was possible without huge inventories of prebuilt stock in all models. It also meant that one production line (or factory) could do the work of several.

The development of just-in-time production required more than the kanban, a point lost on many Westerners. JIT sprang from the understanding of the seven wastes and the need to eliminate them. The key elimination (almost) of material and parts inventories dictated the requirement for reliability and predictability of the plant's machinery and processes. This led to total productive maintenance and made necessary the use of statistical process control and continuous improvement.

With the customer as the driver of production, the control technique for production changed from push to pull, and kanban was introduced as the controlling system. The requirement for small lot sizes, both for elimination of waste and for responsiveness and investment economy, led to the effort to reduce setup time. With all of these factors in place, JIT was born. Without doubt, JIT is the manufacturing system for today. It is adaptable to operations both large and small, high-volume/low-variety, and low-volume/high-variety, and anything in between. In JIT, costs, lead time, and cycle time are reduced, quality is improved constantly, and both customers and the producers and their employees benefit.

RELATIONSHIP OF JIT TO TOTAL QUALITY AND WORLD-CLASS MANUFACTURING

The traditional production line pushes product from the front of the line to the final output, and even to the customers, whereas kanban is the controlling agent in a pull system. The two are incompatible. Similarly, implementing JIT in the absence of a comprehensive total quality system that includes the entire organization can be a problem. The traditional organization is incompatible with JIT, just as the traditional push production system is incompatible with kanban. In a typical manufacturing company, separate departments exist for engineering, manufacturing, purchasing, accounting, and so on, each with distinct boundaries and agendas. JIT is no respecter of boundaries. It requires all departments to respond to its needs. If the manufacturing department has embraced JIT, but the organization as a whole has not at least started a total quality effort, manufacturing personnel will soon encounter obstacles. More often than not there will be outright resistance because JIT's requirements represent change, and departments without a commitment to change will fight it at every step.

As an example, in the defense industry it is common to defray overhead expenses (buildings, utilities, indirect employees' salaries, all fringe benefits, and others) against direct labor dollars as a means of allocating the overhead burden across all contract programs. The more direct labor on a program, the more of the overhead cost accrues to that program. *Direct labor* is defined as the manufacturing, engineering, purchasing, and other labor charged to specific contract programs. The company may also have more than one pool for overhead defrayment, such as a manufacturing pool and an engineering pool. Virtually all of these companies, and the U.S. Department of Defense, pay a great deal of attention to what they call *overhead rate*. In a typical company in the defense industry, overhead rate is calculated by dividing overhead (indirect) expenses by direct labor cost.

Suppose that for an accounting period there were indirect expenses of $200,000. At the same time, the wages paid for direct labor amounted to $100,000. The overhead rate for the period is $200,000/$100,000 = 200\%$. Assume that we had been operating with that 200% rate for some time, and suddenly the manufacturing department discovered JIT. After the period of time necessary for the implementation to start showing results, manufacturing finds that it can eliminate direct labor positions for production control and material control and also use fewer assemblers on the production floor to get the same number of units out the door each period. A typical early change in the direct labor content of the work is 30%–35%. The next period's overhead expense is almost the same, decreasing slightly for removal of fringe benefits for the employees no longer needed, say, to $188,000. The direct labor is down by one-third to $67,000. This yields an overhead rate of $188,000/$67,000 = 281%. That kind of an increase in overhead rate, if sustained, will cause the head of manufacturing serious problems. The accounting department uses this overhead rate as proof that JIT doesn't work. All too often the accounting department blocks further progress in JIT. One might ask, "But isn't that valid if the overhead rate went out of control?" The answer is nobody should care about the overhead rate. It is simply the ratio of two numbers and carries no meaning without a thorough understanding of the two. What happened to the cost of goods sold in this example? Look at the numbers before and after JIT:

	Before JIT	*After JIT*
Indirect Expense	$200,000	$188,000
Direct Labor	100,000	67,000
Materials[8]	500,000	500,000
G&A[8]	50,000	50,000
Cost of Goods Sold	$850,000	$805,000

In this example, it cost the company $45,000 less to produce the same goods after JIT implementation than it did before. Assuming the goods were sold for the same price, that $45,000 becomes pure profit. In the next competition for contracts, the lower cost becomes a competitive advantage (price to the customer can be lowered).

The solution to the overhead rate problem is to change from the obsolete accounting system and adopt an activity-based accounting system or some other more sensible system. In a total quality company, the accounting department is part of the team and would respond to the needs of a production system (JIT) that is actually improving company performance. But if the company as a whole is not involved in total quality, the accounting department, with its own walls and agendas, can be a formidable obstacle to progress. The same is true of other departments on whom manufacturing depends. This example could just as easily have been one involving the engineering department and a design philosophy called *concurrent engineering*. Concurrent engineering requires that from the beginning of a new product's design, manufacturing and other departments (and even suppliers) are directly involved with engineering to make sure, among other things, that the product can be manufactured efficiently when it finally goes into production. Traditional engineering departments do not like to have this kind of help from outsiders and will resist—but not in a total quality setting, where the departments all work for the common goal.

For JIT to bring about the benefits inherent in its philosophy, it must be part of a total quality system. To bring JIT into a company not otherwise engaged in total quality can be worthwhile (and may even enlighten the leadership), but it will be much more difficult, and its results severely restricted.

BENEFITS OF JIT

A discussion of the benefits of JIT must include four very important topics: inventory and work in process, cycle time, continuous improvement, and elimination of waste. The discussion could be expanded to include such topics as reduced time-to-market, improvement of employee work life, flexibility, and employee ownership. All of these are definite benefits of JIT, but this discussion will be confined to the critical four mentioned. These are the usual targets of a JIT implementation.

Inventory and Work-in-Process

Just-in-time attempts to drive inventory to zero. But remember that this is a philosophical objective—an aiming point, if you will. In reality, zero inventory makes no sense. Without some inventory, you have nothing from which to produce your goods. The real objective is to minimize the inventory to the maximum possible extent without shutting down production. It is also important to recognize that there are at least three kinds of inventory. First there is the inventory of raw materials and parts needed to make the product. Traditionally these have filled warehouses, with enough on hand for several weeks of production, or longer. Second, there is the work-in-process (WIP) inventory of semifinished goods. WIP includes all materials and parts that have been put into the production system, including the various stages from the first process to the last within the factory. WIP may be at a workstation undergoing one of the value-adding production processes, or it may be in storage between processes. In a mass production plant, the stored WIP can be substantial. Job shops—the non-mass production, low-volume, high-variety shops—are notorious for their WIP inventory. Third, there is the finished goods inventory. These finished goods are ready for customers, but the customers are not ready for them. Therefore, they are typically stored in warehouses, although some (most notably automobiles) must be stored in yards, unprotected from the elements.

One might ask, what is wrong with inventory? Having materials on hand allows you to produce without worrying about on-time material deliveries. Lots of WIP lets the assembly lines continue when a machine breakdown or some other problem occurs. Having an inventory of stored finished goods means that you can be responsive to customers. If those are positives (and we'll come back to that in a minute), there are also negatives. First, there are the costs of inventorying raw materials and parts, and finished goods. These are the costs of the materials and goods; the labor costs for storage, handling, and protection of the materials and goods; and the cost of warehouses, real estate, and capital equipment used in the inventorying of the materials and goods. Second, there is the cost of spoilage while in inventory. Spoilage can be due to damage, deterioration, corrosion, obsolescence, and so on. Third, there is the cost of taxes. While the product is in inventory, the manufacturer owns it, it has value, and the various governments want their share in the form of taxes.

Now go back to the suggestion made earlier that the three positives associated with inventory might not be so positive after all. The costs discussed earlier are all tangible costs. There are also intangible costs that while difficult to measure precisely are nevertheless significant. Foremost among the intangibles is the fact that as long as the manufacturer holds inventory of materials and WIP at high levels, it is not solving the problems and making the continuous improvements that can bring efficiency. The very presence of these inventories masks the problems, so they go unnoticed and unresolved—being repeated over and over, consuming unnecessary labor, and preventing product quality improvement. Unmasking the production system's problems through the elimination of inventories is a major strength of JIT. Many North American and European companies still tend to see the elimination of inventories as a generator of problems. In reality, the problems are already there, and they are costing a great deal in terms of money and quality, but they are just not apparent with big inventories. Through inventories maintained, tons of money is spent, but no value is added, and needed improvements are not made in the production processes. The inevitable net result is loss of competitive position and market share as enlightened competitors use JIT and total quality to improve their positions.

If a plant could get its production processes under control to the point that they could be relied on to perform as intended, it would be logical to reduce WIP and material/part inventories. However, until the processes are well understood and in control, reducing inventories substantially will certainly result in production stoppages. One philosophy of reducing WIP and lot sizes is to do so in steps. By incrementally lowering WIP and lot sizes, the problems become apparent in a gradual, manageable stream rather than in a torrent, and they can be dealt with. Once through that process, the next logical step is to work with suppliers to deliver materials and parts in smaller, more frequent lots, until finally there is no need for warehousing at all. This clearly requires that the production processes be capable and reliable and that the suppliers be similarly capable and reliable.

This leaves only the finished goods inventory. As the processes and suppliers become more proficient, and the JIT line takes hold, production will be geared to customer demand rather than to sales forecasts. The ability of the JIT line to respond quickly to customer requirements means that it is no longer necessary to store finished goods. The only stored goods should be those in the distribution system, and that level will typically be far less than has been the case under mass production.

JIT strives for zero inventory of any kind. Achieving zero inventory is not a realistic intent, but by aiming at zero and continually reducing inventories, not only do manufacturers cut costs by significant numbers, but the whole continuous improvement process comes to life, resulting in even more savings and improved product quality.

Cycle Time

Production cycle time is defined as the period bounded by the time materials are sent to the manufacturing floor for the making of a product and the time the finished goods are dispatched from the manufacturing floor to a customer or to finished goods storage. Generally speaking, the shorter the production cycle time, the lower the production cost. That may be reason enough to pay attention to cycle time, but there are other benefits. Short cycles improve a factory's ability to respond quickly to changing customer demands. The less time a product spends in the production cycle, the less chance there is for damage.

We are accustomed to thinking of a mass production line as having the shortest of cycle times, and there have been startling examples of this. Henry Ford's Model T lines (up to 2,000,000 per year, all the same, all black) achieved remarkable cycle times even by today's standards. For example, Ford's River Rouge facility took iron ore in the front door and shipped completed cars out the back door in four days.[9] When one considers that the Ford cycle included making the steel, in addition to stamping, casting, machining, and assembly, it is all the more amazing. One of his secrets was no variability in the product. Modern lines have the complication of different models and virtually unlimited options.

A modern auto assembly line cannot be compared with Ford's Model T line because the complexity and variability of the contemporary car is so much greater. However, the best lines beat Ford's cycle time for assembly. The differences in JIT lines and mass production lines is substantial. For example, comparisons between JIT plants and traditional mass production plants reveal that JIT plants can assemble automobiles in 52% of the time it takes traditional plants. Because there is very little waiting in a JIT line, one can assume the cycle time is one half of that for traditional lines. Interestingly, though not directly related to cycle time, traditional lines produce three times as many defects and require nearly twice the factory space. In addition, JIT plants can operate with a 2-hour parts inventory, while traditional plants typically need a 2-week supply.[10]

Consider the following example that helps bridge the issues of inventory and cycle time. The product was a line of very expensive military avionics test systems. The factories (two) were rather typical electronics job shops. Before being converted to JIT, they were struggling with a production schedule requiring the assembly of 75 large, complex printed circuit boards per day. They rarely met the goal, usually achieving about 50. The attempted solution involved pushing more parts into the front end of the assembly process, hoping that would force more out the other end as finished, tested boards. The computer system revealed that, at any point in time, about 3,500 boards were in the process. At the rate of 50 completed boards per day, and 3,500 boards in WIP, simple arithmetic showed that the cycle time for the average board was 13 weeks. Common sense said that 13 weeks was much too long for assembling these boards, but checking with others in the industry revealed that this was a typical cycle time. The company also found that it made absolutely no difference in final output rate to force more materials into the front of the process. This merely increased the number of boards in WIP.

With a production rate of 50 boards a day and 3,500 boards in process, one can imagine the difficulty in keeping track of where the boards were, scheduling them into and out of the various processes, storing, retrieving, and safeguarding. Such tasks were nearly impossible. More than 100 people were charged with handling and tracking the boards, adding no value whatever to the product. Further, because the assemblers were being pushed to their limits, quality suffered. The net result was that nearly half of total direct labor was spent repairing defects. That is not adding value either. However, once again, checking with other manufacturers revealed that this was typical. A critical factor was that customer delivery schedules could not be met unless a solution was found. Initially, the company had to subcontract a great many boards, but that was a work-around, not a solution.

The eventual solution was to implement JIT techniques on the production floor. After a couple of quick pilot runs, in which it was discovered that the most difficult of the boards could be assembled and tested in 8 days (versus 13 weeks), management was convinced, and JIT was implemented at both plants, following the WIP reduction and lot size scheme outlined in the previous section. In very short order, the board cycle time fell to about 5 days, and board quality improved dramatically. That enabled the company to eliminate the 100-plus positions that had handled the boards and eventually many other non-value-adding positions as well. The system delivery on-time rate went to 98% (unheard of for this kind of product), customer satisfaction improved, and a respectable profit was made.

The thing to remember about cycle time is this: any time above that which is directly required by the manufacturing process is not adding value and is costing money. For example, assume we use two processes to manufacture a product, and the total time consumed within the processes is 2 hours. It is determined that the actual cycle time is 3 hours. That means that 2 hours of the cycle is adding value and the other hour is not. Invariably this means a bottleneck is preventing the product from flowing from one process directly into the next without delay. The key is to detect the bottleneck and do something about it. It may be that a plant procedure requires inspection, logging, and a computer data entry. Are these tasks really necessary? Can they be eliminated? If they are necessary, can they be streamlined?

The extra hour may be the result of a problem in one of the processes. For example, it may be that the second process is no longer 1 hour in duration but 2. If the latter is the case, in a traditional production plant, the product flowing out of the first process will stack up at the input of the second process, because process 1 will continue to crank out its product at the rate of one unit per hour—whether process 2 is ready for it or not (see Figure 14–7). The surplus product at the input to process 2 will have to be stored for safety and housekeeping reasons, thus obscuring the fact that there is a problem.

As long as the problem persists, WIP will build, output will stay at one unit every 2 hours, but cycle time will increase as backlog builds up in front of process 2: the first unit went through the production system in 3 hours, and one unit per hour was expected after that, but the process is actually achieving one unit every 2 hours. Cycle time increases by 1 hour for each piece—for example, 8 hours later the sixth unit into process 1 will come out of process 2. Such an imbalance would not escape notice for long, and it would be corrected, but by then several pieces of WIP would be between the processes.

Figure 14–7
Cycle Time Example

Piece #	Process 1 In	Process 1 Out	Wait Time (In Hours)	Process 2 In	Process 2 Out	Cycle Time (In Hours)
1	7 a.m.	8 a.m.	0	8 a.m.	10 a.m.	3
2	8 a.m.	9 a.m.	1	10 a.m.	12 noon	4
3	9 a.m.	10 a.m.	2	12 noon	2 p.m.	5
4	10 a.m.	11 a.m.	3	2 p.m.	4 p.m.	6
5	11 a.m.	12 noon	4	4 p.m.	6 p.m.	7
6	12 noon	1 p.m.	5	6 p.m.	8 p.m.	8
7	1 p.m.	2 p.m.	6	8 p.m.	9 p.m.	8
8	2 p.m.	3 p.m.	6	9 p.m.	10 p.m.	8
9	3 p.m.	4 p.m.	6	10 p.m.	11 p.m.	8
10	4 p.m.	5 p.m.	6	11 p.m.	12 midn.	8
11	5 p.m.	6 p.m.	6	12 midn.	1 a.m.	8
12	6 p.m.	7 p.m.	6	1 a.m.	2 a.m.	8
13	7 p.m.	8 p.m.	6	2 a.m.	3 a.m.	8

Suppose that the problem in the second process was corrected as the sixth unit was completed. Everything is back to the original 2-hour process time, but by now there are seven more units through process 1, on which the cycle time clock has already started. If stable from this point forward, the cycle time will remain at 8 hours. We started with a process that had 2 hours of value-adding work and a 3-hour cycle. We now have a 2-hour value-adding process time and an 8-hour cycle. If some means is not taken to cause the second process to catch up, every time there is a glitch in process 2, the cycle time will grow. In a traditional plant, with literally dozens of processes, such conditions could go on forever. As observed earlier, some would hold that having the seven units from the first process sitting on the shelf means that process 1 could be down for a complete shift without causing a problem for the second process—it would merely draw from the seven.

In a JIT plant, the situation described here would never happen. Process 1 would not produce an additional piece until process 2 asked for it (kanban). At the start, process 1 produces one unit to enable process 2. When process 2 withdraws it, process 1 is signaled to produce another. If for any reason, when process 1 completes its second unit, process 2 is not ready to withdraw it, process 1 goes idle. It will stay idle until signaled to produce another—be it a few minutes or a week. No WIP inventory is produced. By process 1 going idle, alarms go off, quickly letting the appropriate people know that something has gone wrong. If the second process has a problem consuming too much time, it gets attention immediately. Similarly, if there is a delay getting the output of the first process to the second because of an administrative procedure, that too will be dealt with quickly, because it will cause problems throughout the overall process until it is solved.

Any contributor to cycle time is apparent in a JIT environment, and JIT philosophy calls for continuous improvement and refinement. Wait time in storage is simply not a factor in JIT, because nothing is produced in advance of its need by the succeeding process. That single factor can easily remove 80%–90% of the cycle time in a traditional factory. In the earlier example of the printed circuit board factories, the initial reduction of cycle time from 13 weeks (65 working days) to 8 days was simply the elimination of storage time. That was a reduction of 88%. Further refinement, made possible because of the visibility afforded by JIT, brought the cycle to 4 days, or only 6% of the original cycle. Taking it further was restricted by procedural and governmental requirements. In a commercial setting, however, the same boards could probably have been produced in a 2-day cycle with no new capital equipment.

Before JIT, manufacturers tried to cut cycle time with automation. But that was not the answer. The solution was found in better control of production, and that was obtained with JIT. JIT is the most powerful concept available for reducing cycle time.

Continuous Improvement

Continuous improvement has been discussed in several other chapters and sections of this book. By now, you should have a good understanding of its meaning as applied in a total quality context. Continuous improvement seeks to eliminate waste in all forms, improve quality of products and services, and improve customer responsiveness—and do all of this while at the same time reducing costs. Perhaps a note of caution should be added in regard to interpretation of what constitutes improvement. Problem solving is not necessarily improvement. If a process that had previously been capable of producing 95 out of 100 good parts deteriorates to a level of 50 good parts, and the problem is found and corrected to bring the process back to where it had been, that is maintenance, not improvement. *Maintenance* is restoring a capability that previously existed. On the other hand, if a process was capable of 95 good parts out of 100 produced, and a team developed a way to change the process to produce 99 good parts, that would be improvement. It is important to differentiate between maintenance and improvement. Maintenance is important, and it must go on, but in the final analysis, you end up where you started. *Improvement* means becoming better than when you started. Continuous improvement is to repeat that improvement cycle, in W. Edwards Deming's words, constantly and forever.

The discussion of continuous improvement in this chapter is to explain how JIT supports continuous improvement. The traditional factory effectively hides its information through inventories of parts, WIP, and finished goods; people scurrying about, everybody busy, whether any value is being added or not. The JIT factory is visual: its information is there for everyone to see and use. Quality defects become immediately apparent, as do improper production rates—whether too slow or too fast. Either of these, for example, will result in people stopping work. While that is not acceptable behavior in a mass production factory, in a JIT plant it is encouraged and expected.

A true story from Toyota tells of two supervisors, one from the old school and unable to adapt to JIT, and the other ready to try JIT even if it did seem strange.[11] The first supervisor refused to allow his line to be stopped, whereas the second didn't hesitate to stop his. At first the line operated by the second supervisor was producing far fewer cars than the other line, because it was stopping for every little problem. These problems had been common knowledge among the workers but not to the supervisors. The problems were solved one by one as a result of stopping the line for each one. After 3 weeks, the second supervisor's line took the lead for good. The first supervisor believed that stopping the line would decrease efficiency and cost the company money. As it turned out, the reverse was true. By stopping the line to eliminate problems, efficiency and economy were enhanced. The only reason for stopping a line is to improve it, eliminating the need for stopping again for the same reason.

In a mass production plant, the sight of idle workers will draw the ire of supervisors in no uncertain terms. But in a JIT situation, the rule is if there is a problem, stop. Suppose that a preceding process has responded to a kanban and provided a part to a succeeding process. The succeeding process finds that the part is not acceptable for some reason (fit, finish, improper model, or something else). The succeeding process worker immediately stops, reporting the problem to the preceding process and to supervision. Perhaps an *andon* (a Japanese word meaning "lamp") signal will be illuminated to call attention to the fact that his process is shut down. The problem is to be solved before any more work is done by the two processes, which means that downstream processes may soon stop as well, because their demands through kanban cannot be honored until the problem is fixed and the processes are once again running. This is high visibility, and it is guaranteed to get the proper attention not only to solve the immediate problem but to improve the process to make sure it does not happen again.

Consider the following example. A few weeks after JIT implementation was started in a New York electronics plant, there was a line shutdown. At the end of this line was a test station that was to do a comprehensive functional test of the product. There was an assembly all set up for test, but the technician had stopped. The line's andon light was illuminated. A small crowd gathered. The problem was that the test instructions were out-of-date. Over time the test instruction document had been red-lined with changes and had, up until that point, been used without apparent difficulty. But a company procedure required that any red-lined document be reissued to incorporate the approved changes within 1 year of the first red-line. The 1-year clock had expired months earlier, and the technician, with guidance from quality assurance, properly stopped testing. When management asked why the document had not long since been updated, it was found that the documents seldom were updated until the entire job was completed. In many cases, jobs lasted several years. Holding all formal revisions until a job was completed meant that documentation was revised just once, thereby saving considerable expense. Of course, in the meantime, manufacturing was using out-of-date or questionable information. The standard work-around seemed to be that when a system couldn't be completed for delivery, waivers were generated, allowing the tests to be conducted with the outdated red-lined procedures. This had been going on for years but never became apparent to the levels in manufacturing and engineering that could solve it. In this case, it took about 20 minutes to have the problem solved. Without JIT to highlight it, the problem would, in all probability, still exist.

What had happened because of JIT was a stop at the test station. That also shut off kanbans through the preceding processes. In short order the line stopped, getting the attention needed to eliminate the problem. If the plant had been operating in the traditional (non-JIT) way, the assemblies would have piled up at the test station for a while and then the production control people would have carted them off to a work-in-process storage area—out of sight. Eventually the inventory of previously tested assemblies would be consumed, and there would have been a brushfire from which a procedural waiver would emerge to enable the test technician to pull the untested assemblies from WIP stores and quickly get them tested so system deliveries could be made. And it would have repeated time and again, just as it had been doing surreptitiously in the past.

This is not an uncommon scenario. Fundamentally, it is the result of departments not communicating. Engineering is trying to save money by reducing the number of documentation revisions. Meanwhile, manufacturing may be producing obsolete and unusable product because their documentation is not up to date. At best, it results in the continual firefighting that saps the collective energy of the organization, leading to quick-fix, work-around "solutions" that let you get today's product out but only making each succeeding day that much more difficult. JIT, by highlighting problems, is quick to dispel the quick-fix mentality, demanding instead that problems be eliminated for today and tomorrow and forever.

A better illustration of JIT's ability to reveal your real problems is the analogy of the lake (see Figure 14–8). You look out over a lake and see the calm, flat surface of the water and perhaps an island or two. From this observation, you conclude that the lake is navigable, so you put your boat in and cast off. You avoid running into the islands because they can be seen plainly and there is plenty of room to steer around them. However, a rock just below the surface is not evident until you crash into it. It turns out there are lots of rocks at various depths, but you can't see them until it is too late. This is like a traditional factory. The rocks represent problems that will wreak havoc on production (the boat). The water represents all the inventory maintained, raw materials and parts, WIP inventory, even finished goods. Now if you make the change to just-in-time, you start reducing those inventories. Every time you remove some, the level of the water in the lake is lowered, revealing problems that had been there all along but that were not eliminated because they couldn't be seen. You just kept running your boat into them, making repairs and sailing on to the next encounter. But with the lower water level, the problems become visible and can be eliminated. Clear sailing? Probably not. Other rocks are no doubt just below the new lower surface level, so you have

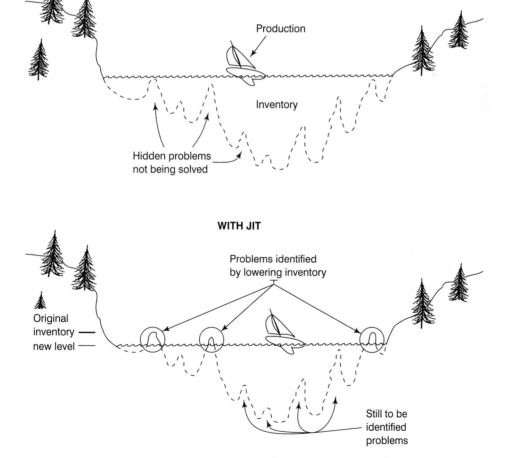

Figure 14–8
JIT Exposes Hidden Problems

to take some more water out of the lake (remove more inventory), enabling you to identify and eliminate them. Like most analogies, our lake doesn't hold all the way to the logical conclusion of zero inventory, because the lake would be dry by then. But remember, true zero inventory doesn't hold either. As was said before, it is a target to aim at but never to be fully reached.

JIT is by nature a visible process, making problems and opportunities for improvement obvious. Moreover, when problems do occur in a JIT setting, they must be solved and not merely patched up, or they will immediately reappear. Visibility to all levels, from the workers to the top executive, means that the power to make necessary changes to eliminate problems and improve processes is available.

Elimination of Waste

In the preceding three sections it was shown how just-in-time facilitates reduction of inventories and cycle time and promotes continuous improvement. This section will show that JIT is also a powerful eliminator of waste. As discussed earlier in this chapter (in the "Rationale for JIT" section), Taiichi Ohno created the JIT system specifically to eliminate the seven wastes he saw in the Ford mass production system. Let's look at them one at a time and see how JIT helps in their elimination.

1. *Waste arising from overproducing.* Mass production pushes materials into the front of the factory in response to market forecasts. These raw materials are converted to finished goods and pushed through the distribution system. The first real customer input into the process is at the retail level. If customers don't want the goods, they will eventually be sold at much lower than anticipated prices, often below their actual cost. That is waste to the producer. In addition, producing goods for which there is not a matching demand is a waste to society by using resources to no purpose. In a JIT environment, the customers enter the system at the beginning, pulling goods from the distribution system and in turn from the manufacturer. The JIT factory produces nothing without a kanban, which, in effect, originates with a customer.

The same is true within the two kinds of factories. A fender stamping press in a mass production factory will continue to stamp out fenders even though the final assembly line, which uses the fenders, is stopped. The overproduction must then be handled by people who contribute nothing to the value of the product, stored in buildings that would otherwise be unnecessary, tracked by people and systems that do nothing but cost money. In a JIT factory, the fender stamping press will shut down unless it receives kanbans requesting more fenders. No overproduction. Of all the wastes, overproduction is the most insidious because it gives rise to all the other types of waste.

2. *Waste arising from waiting* (*time*). Wait time can come from many causes: Waiting for parts to be retrieved from a storage location. Waiting for a tool to be replaced. Waiting for a machine to be repaired or to be set up for a different product. Waiting for the next unit to move down the line. JIT parts are typically located at the workstation, not in some central staging area or warehouse. JIT sets time aside for tool and machine maintenance, so replacement or repair during a production period is rare. Whereas setup times for machines in mass production plants tend to take hours (or even more), JIT factories devote a great deal of attention to setup time, typically reducing them to a very few minutes. In a typical factory, an operator is assigned to each machine. While the machine is running under automatic control, the operator has nothing to do but wait. In a JIT factory, the same operator may run five machines, arranged so that he or she can easily see and control all five without much movement. As three machines are running automatically, the operator may set up a fourth and unload the fifth, for example. In this way, the day is no longer mostly wait time.

Perhaps the biggest waste associated with waiting involves not human waiting but inventory waiting. In the traditional setting, raw materials and parts can sit idle for weeks and months before they are needed. Work in process may wait weeks to have a few hours of value-adding work done. Finished goods may wait very long periods for customers. JIT does not allow any of these waits to occur, and the carrying expense is eliminated.

3. *Waste arising from transport.* Mass production factories tend to buy their materials and parts in very large quantities from the lowest price (as opposed to lowest cost or best value) source, regardless of the distance from the source to the factory. JIT factories of necessity must buy in small quantities (no warehousing) with frequent deliveries, often several times a day. That means that the suppliers should be relatively close to the factory, cutting transportation.

Transportation within plants can be a very big-cost item too. Moving things costs money and time and increases exposure to damage. Moving materials in and out of storage areas, to and from the floor, back and forth across the factory from process to process is waste. None of that happens with JIT. Production materials are delivered to the point of use in a JIT factory, so they are not shuttled in and out of storage or put in temporary storage to be moved again before use. Factories are arranged to minimize distances between adjacent processes, whereas the same product manufactured in the traditional factory could log thousands of feet of movement before completion.

4. *Waste arising from processing itself.* Any process that does not operate smoothly as intended but instead requires extra work or attention by the operator is wasteful. An example is the necessity for the operator to override an automatic machine function to prevent defective product. Because one of the basic tenets of JIT is continuous improvement of processes, wasteful processes are soon identified and improved to eliminate the waste. That is far more difficult in the traditional production environment because of its emphasis on output, not process improvement.

5. *Waste arising from unnecessary stock on hand.* Any stock on hand has storage costs associated with it. When that stock is unnecessary, the costs are pure waste. Included in these costs are real estate, buildings, employees not otherwise needed, and the costs of tracking and administration. Because JIT attempts to eliminate stock, stock that is not necessary is just not tolerated.

6. *Waste arising from unnecessary motion.* JIT plants are laid out to minimize motion of both workers and product. Motion takes time, adds no value, makes necessary additional workers, and hides waste. The contrast between a JIT plant laid out with product orientation and the traditional plant laid out with process orientation is profound (see Figure 14–9). In the traditional plant, there is much motion, with people and product shuttling all over the place. In a JIT plant, motion is almost undetectable to a casual observer.

7. *Waste arising from producing defective goods.* Defective goods will surely cost money in one of three ways: (a) the product may be reworked to correct the deficiency, in which case the rework labor and material costs represent waste; (b) it may be scrapped, in which case the cost of the materials and the value added by labor has been wasted; or (c) it may be sold to customers who, on discovering that the product is defective, return it for repair under warranty and may be dissatisfied to the extent they will never buy this manufacturer's products again. Warranty costs represent a waste, and the potential for a lost customer is great, portending a future loss of sales.

In a traditional factory, it is possible to produce large quantities of product before defects are discovered and the line corrected. It is not uncommon in mass production for a company to keep the line running intentionally, producing defective product, while trying to figure out what has happened and devising a solution. It is considered less troublesome to fix the defective product later than to shut down the line. In JIT, however, because line stops are anticipated, and because the preferred lot size is one unit, it is improbable that more than one defective unit could be produced before shutting down the line.

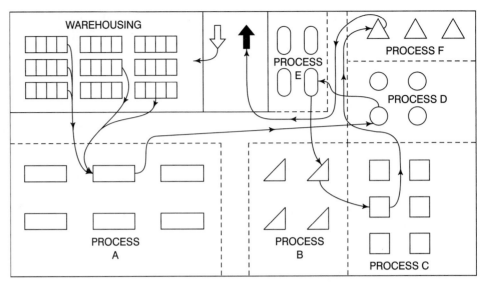

Traditional Factory Organized by Processes
(illustrating process control for one product)

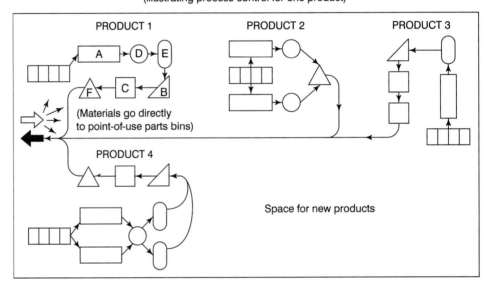

JIT Factory Organized by Product
(illustrating process flow for four products)

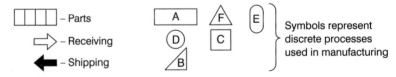

Figure 14–9
Comparison of Factory Floor Layouts: Traditional versus JIT

Dr. M. Scott Myers, author of the landmark book *Every Employee a Manager,* made the case for an eighth waste: the waste arising from the underutilization of talent. Myers believed that the most damaging of the eight wastes is the waste of talent.[12] If all the talents of all employees were brought to bear on the problems and issues of production, the other wastes would probably disappear. This is the rationale for both employee involvement and teamwork. JIT is designed to make use of the ideas and talents of all employees through team activities and employee involvement, in an environment that fosters the open and free interchange of ideas, all of which are foreign to the traditional production systems. Elimination of waste is an integral focus of just-in-time by design. No other production system looks at waste except after the fact.

REQUIREMENTS OF JIT

For a factory to operate as a just-in-time production facility, a number of steps must be taken. It is very important that JIT implementation be a part of a larger total quality program; otherwise, many interdepartmental roadblocks will crop up as time passes. Like total quality, JIT requires an unwavering commitment from the top, because production is more than just the manufacturing department. If these two elements are in place (a total quality program and a commitment from the top), then JIT implementation should be within reach. The following discussion touches on the issues that must be addressed as the implementation progresses.

Factory Organization

The JIT plant is laid out quite differently from what most people are accustomed to. Most non-JIT factories are set up according to the processes that are used. For example, there may be a welding shop, a machine shop, a cable assembly area, a printed circuit board assembly area, a soldering area, and so on. Each of these discrete processes may be set up in separate parts of the factory (all machining operations done in the machine shop, all cable assembly done in the common cable and harness area), no matter which of many products it might be for (refer to Figure 14–9). The JIT plant attempts to set up the factory by product rather than process. All the necessary processes for a given product should be colocated in a single area and laid out in as compact a manner as possible.

The chart at the top of Figure 14–9 represents the old process-oriented factory. Each of the processes has its own territory within the plant. Additionally, an area is dedicated to warehousing that is used for storage of production materials, work-in-process that is waiting for the next process, and perhaps finished goods awaiting orders. There is also an area set aside for shipping and receiving. Materials are received, inspected, processed, and sent to the warehouse area. Finished goods are taken from the warehouse or from final assembly, packed, and shipped. The upper illustration in Figure 14–9 maps the movement from warehouse through the processes and finally to shipping in a traditional factory.

The lower illustration in Figure 14–9 represents a JIT factory that is set up to manufacture four different products. The warehousing area is gone. This cannot happen overnight, but an objective of JIT is to eliminate all inventories. The second thing to notice is that the factory is divided into discrete areas dedicated to the different products rather than to processes. Each product area is equipped with the processes required for that product. Parts bins are located right in the work area. These parts bins may have from a few hours' supply to a month or more, depending on the degree of maturity of the JIT implementation and the nature of the product and its anticipated production life.

Mapped out is a typical work-flow diagram for one product in the upper illustration of Figure 14–9. Parts and materials are pulled from several locations in the warehousing area, and moved to a Process A workstation. These materials may be in kit form (all the parts needed to make one lot of a product). The work instructions call for process A first, followed by process D. If process D is busy when the lot is finished by process A, the lot, now WIP, may be stacked up in a queue at process D or taken back to the warehouse for safekeeping. Eventually process D will process the kit and it will be sent to Process E, perhaps waiting in queue or in the warehouse. This same sequence is repeated through process B, process C, and process F. From there it goes to shipping. The diagram does not show any trips back to the warehouse between processes, but that could very well happen after every step. The movement flow diagram represents a best-case scenario. This was done purposely to ensure clarity.

Now observe the flow in the JIT Factory of Figure 14–9. Product 1 is set up to follow exactly the same processing sequence (from parts bins to Process A, then process D, process E, process B, process C, process F, and to shipping). In this case, the parts come straight from the bins located in the work cell, not from the warehouse, and not in kit form, which is a waste of effort. The work cell is laid out in a U shape for compactness, to keep all the work cell members close to each other. The WIP flows directly from process to process without a lot of wasted movement. Moreover, because this is a JIT work cell, there will be small lot sizes, with work pulled through the process sequence by kanban. That means there will be no queue time on the floor or in the warehouse. Cycle time for this product in the JIT work cell can be expected to be less than half of that for the same product in the factory at the top of Figure 14–9. An 80%–90% reduction would not be unusual.

Before one can lay out a JIT factory, the processes required for the product must be known. This is usually not a problem. Typically, the most difficulty comes in determining how much of a process is needed. How many minutes of a process does the product use? One would think that if the product had been built before in the traditional way, one should know how much process time is required at each step. This may be a starting point, but typically it is not very accurate. With all the wasted motion and waiting time in queues and in the warehouse, the real processing time becomes obscured. However, you can use the best information available and refine it over time. Now that the processes are put right into the product work cell, having just the right amount is important.

In the case of product 4 in Figure 14–9's JIT factory, it was determined that the product required more capability in process A and process E than was available from single workstations, so they were doubled. Suppose that a product flow of 120 units per hour is anticipated. Each process has the following estimated capability for this product:

A: 75 units/hour   B: 150 units/hour   C: 130 units/hour
D: 120 units/hour   E: 70 units/hour   F: 135 units/hour

Because processes A and E are estimated to be capable of only about 60% of the anticipated demand, there is no point in trying to improve them. Rather, the process capability was doubled by putting in parallel equipment/workstations. This is a beginning. We now can watch for excess capacity that can be removed from the work cell, or for bottlenecks that require other adjustment. Work cells are coarsely tuned at first, with fine-tuning taking place during the initial runs. Excess capacity should be removed, just as required added capacity must be brought into the work cell. Bottlenecks will be quickly discovered and corrected. From there on, it is a matter of continuous improvement to increase efficiency forever.

Training/Teams/Skills

Assuming an existing factory is converted to just-in-time, one would assume that the people who had been operating it would be capable of doing it under JIT. Naturally, many of the skills and much of the training necessary for the traditional factory are required under JIT, but JIT does require additional training. First, the transition from the traditional way of doing things in a factory to JIT involves profound changes. It will seem that everything has been turned upside down for a while. People should not be exposed to that kind of change without preparation. It is advisable to provide employees with training about why the change is being made, how JIT works, what to expect, and how JIT will affect them. Initial training should be aimed at orientation and familiarization. Detailed training on subjects such as kanban, process improvement, and statistical tools should be provided when they are needed—a sort of just-in-time approach to training.

Most factory workers are accustomed to working individually. That will change under JIT, which is designed around teams. A JIT work cell forms a natural team. The team is responsible for the total product, from the first production process to the shipping dock. Perhaps for the first time the workers will be able to identify with a product, something that they create, and the processes they own. This doesn't happen in a traditional factory. But with JIT, it is important to understand that workers must function as a team. Each will have his or her special tasks, but they work together, supporting each other, solving problems, checking work, helping out wherever they can. This may require some coaching and facilitating.

It was enough in the old way of production that workers had the skills for their individual processes. They did not need additional skills because they were locked into one process. This is not the case with JIT. Specialists are of far less value than generalists. Cross training is required to develop new skills. As a minimum, work cell members should develop skills in all the processes required by their product. Naturally, there are limits to this. We do not propose that all the members of a work cell become electronics technicians if their cell employs one for testing the product, but the cross training should broaden their skills as far as is reasonable. Even on the issue of technical skills, it is beneficial to move in that direction. For example, if an operator's task is to assemble an electronic assembly that will be part of an end item device, there is no reason that operators couldn't test it when they complete the assembly. Go/no go testers can be built to facilitate testing any electronic assembly, and they can be simple enough to operate that the assembler can easily perform the test. This frees the technician for the more complicated tests downstream and ensures that the assembly is working before it is passed on to the next higher level. It also gives operators a sense of ownership and accomplishment. Over time they may even be able to troubleshoot an assembly that fails the test.

Requiring multiple skills in JIT teams is important for several reasons. First, when a team member is absent, the work cell can still function. Second, problem solving and continuous improvement are enhanced by having more than one expert on whatever process is in question. New people will have fresh new ideas. Third, when one of the cell's processes starts falling behind, another member can augment the process until it is back on track.

Establish Flow/Simplify

Ideally, a new line could be set up as a test case to get the flow established, balance the flow, and generally work out initial problems. In the real world, this may not be feasible. Normally the new line is set up to produce deliverable goods. What typically happens is a line is set up, then operated with just a few pieces flowing through to verify the line's parameters. It is very important to maintain strict discipline on the line during pilot runs. Procedures must be strictly adhered to. Each operator must stay in his or her assigned work area, with no helping in another process. Only by pilot runs conducted this way will the information gained be meaningful and valid. It will allow process times to be checked, wait times to be assessed, bottlenecks to be identified, and workers to become synchronized. It is not necessary to have a pull system in place for these preliminary runs, because only a few pieces will be involved. In fact, until the flows have been established, kanban is not possible.

The second thing to look for in these pilot runs is how well the line accommodates the work. Are the workstations positioned for the least motion? Is there sufficient space, but not too much? Can the operators communicate easily with each other? Is the setup logical and simple? Can any changes be made to make it better, simpler? Don't overlook the processes themselves. Ultimately that is where most of the simplification will occur.

Kanban Pull System

Having established the flow and simplified it to the extent possible, the company can now introduce the kanban pull system. As the work cell is being designed, the kanban scheme should be developed. For example, will a single or double kanban card system be used, or kanban squares or bins? Or will some combination or a different variation be used? A kanban plan must be tailored to the application: there is no single, best, universally applicable kanban system.

Readers who are familiar with manufacturing may know that cards have been used in the manufacturing process as long as anyone can remember. They take the form of traveler tags, job orders, route sheets, and so on, but they are not at all the same as kanbans. These cards push materials and parts into a production process, such as PC-board stuffing. When the boards controlled by the card are all stuffed (the electronic components have been inserted into the boards), the entire batch is pushed to the next process. Ready or not, here they come. The next process didn't ask for them and may not be ready for them—in which case, they will stack up in front of the process or be removed from the production floor and stored with other waiting WIP. By contrast, in a JIT line, the succeeding process signals the preceding process by kanban that it needs its output. Be sure to understand the distinction. With kanban, the succeeding process pulls from the preceding (supplying) process. The kanban always tells the supplying process exactly what it wants and how many. The supplying process is not authorized to make more product until the kanban tells it to do so—nothing waiting, no stored WIP.

The Toyota system uses two types of kanbans: the withdrawal kanban and the production kanban. The withdrawal kanban, also called the move kanban, is used to authorize the movement of WIP or materials from one process to another (see Figure 14–10). The kanban will contain information about the part it is authorizing for withdrawal, quantity, identification of containers used, and the two processes involved (supplying and receiving). The production kanban authorizes a process to produce another lot of one or more pieces as specified by the kanban (see Figure 14–11). The kanban

PRODUCING PROCESS HARNESS ASSY <u>BHA-15</u>	LOCATION <u>BHA-15</u> SQ. F 1	PART NO. <u>3371-10130</u>	WITHDRAWING PROCESS PANEL INTEG <u>BPT-1</u>
	CONTAINER TYPE <u>N/A</u>	PART DESCR <u>BETA HARNESS</u>	
	CONTAINER CAPACITY <u>——</u>	NO. WITHDRAWN <u>1</u>	RECEIVING LOC BPT-1 WS

Figure 14–10
Withdrawal Kanban

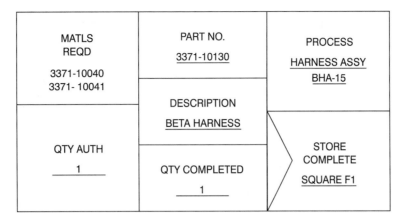

Figure 14–11
Production Kanban

also gives the description of the piece(s) authorized, identifies the materials to be used, designates the producing process workstation, and tells the producing process what to do with it when it is completed.

Consider the operation of two processes in a manufacturing sequence to see how this works in practice. Figure 14–12 shows a preceding process that does grinding on metal parts. This is the supplier for the parts finishing workstation, the succeeding process. Figure 14–12 shows five segments, described in the following paragraphs.

Segment 1 reveals that the finishing workstation has containers at both its In and Out areas. The container at the In area carries a move kanban (MK), and has one part left to be used. The container at the Out area has five finished parts in it and is waiting for the sixth. Back at the grinding workstation the Out container is filled with the six parts authorized by the production kanban (PK) attached. The container at the In area is empty, and work is stopped until another production kanban appears.

Segment 2 shows that the finishing workstation has completed work on the six parts, emptying the container at its In area. The empty container with its attached MK for six parts is taken back to the grinding workstation, which is ready to supply the parts.

Segment 3 shows that when the empty container is received at the grinding workstation, the move kanban is removed from the empty container and attached to the full container, which is sitting at the process's Out area. This authorizes movement of the six parts to the finishing workstation. At the same time, the production kanban was removed from the full container and attached to the empty one, which is placed at the grinding workstation's Out area. This authorizes the grinding process to grind six more pieces.

Segment 4 shows that the finishing process has now processed two parts. The empty container at the In area of the grinding process has been taken back to the preceding process in order to obtain the parts it needs to grind six new pieces.

Segment 5 shows the finishing workstation halfway through its six pieces, with the grinding process started on its next six pieces. This cycle will repeat itself until there is no more demand pull from the right side (from the customers and the final processes).

The finishing workstation had its Out parts pulled by the next process in Segment 2, triggering finishing's pull demand on grinding in Segment 3. That in turn resulted in grinding's pull from its previous process in Segment 4. The pulls flow from the right (customer side), all the way through the production processes to the left (supplier side). When demand stops at the customer side, pulling stops throughout the system, and production ceases. Similarly, increase or decrease in demand at the customer side is reflected by automatically adjusted pulls throughout the system.

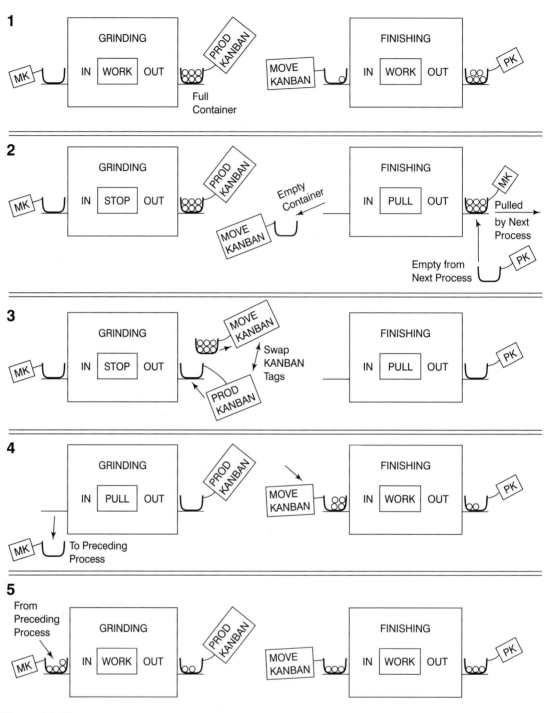

Figure 14–12
Dual Card Kanban System

As suggested earlier, it is not always necessary to use actual kanban cards. In many applications, it is necessary only to use kanban squares, kanban shelves, or kanban containers. In Figure 14–12, for example, the two processes could have used any of these devices. The Out side of the grinding workstation could have the right side of its table top marked out in six kanban squares. One part ready for finishing would be placed on each square, like checkers on a checkerboard. The signal to grind six more parts would be the finishing workstation's taking of the parts, leaving the kanban squares empty. In this case, the empty kanban square is the signal to produce more. Marked-off shelf areas, empty containers designated for so many parts, and various other devices can be used. Combinations are the rule.

Kanban is a shop floor control/management system. As such, it has some rules that must be observed:[13]

Rule 1: Do not send defective product to the subsequent process.   Instead, stop the process, find out why it was made defective, and eliminate the cause. It will be much easier to find the cause immediately after it happened than it would be after time has elapsed and conditions have changed. Attention to the problem will escalate rapidly as subsequent processes come to a halt, forcing resolution. Only then should the subsequent process be supplied.

Rule 2: The subsequent process comes to withdraw only what is needed, when needed.   There can be no withdrawal without a kanban (of some sort). The number of items withdrawn must match the number authorized by the kanban. A kanban must accompany each item.

Rule 3: Produce only the exact quantity withdrawn by the subsequent process. Never produce more than authorized by the kanban. Produce in the sequence the kanban are received (first in, first out).

Rule 4: Smooth the production load.   Production flow should be such that subsequent processes withdraw from preceding processes in regular intervals and quantities. If production has not been equalized (smoothed), then the preceding process will have to have excess capacity (equipment and people) to satisfy the subsequent process. The earlier in the production process, the greater the need for excess capacity. Because excess capacity is waste, it is undesirable. The alternative would be for the processes to "build ahead" in anticipation of demand. This is not allowed by rule 3. Load smoothing will make or break the system, because it is the only way to avoid these two intolerable alternatives.

Rule 5: Adhere to kanban instructions while fine-tuning.   In the previous section, we said that for a kanban system to work, the flow must first be established. Kanban cannot respond to major change, but it is a valuable tool for the fine-tuning process. All the production and transportation instructions dealing with when, how many, where, and so on, are designated on the kanban. If the manufacturing process has not been smoothed, one cannot, for example, tell a preceding process to do something early to compensate. Instructions on the kanban must be observed. Adhering to the kanban's instructions while making small, fine-tuning adjustments will help bring about optimum load smoothing.

Rule 6: Stabilize and rationalize the process.   The processes need to be made capable and stable. Work instructions/methods must be simplified and standardized. All confusion and unreasonableness must be removed from the manufacturing system, or subsequent processes can never be assured of availability of defect-free material when needed, in the quantity needed.

Observing the six rules of kanban all the time is difficult, but it is necessary if the production flow in a JIT system is to mature and costs are to be reduced.

Kanban is often used by itself for shop floor control very effectively, but it can also be used in conjunction with automation, such as bar code and computer augmentation. Computer-based kanban systems exist that permit the fundamental kanban system in a paperless environment. As with automation in general, such a computerized system must be designed to suit the application. Applying technology simply for technology's sake is never a good idea. Whatever you do, it is best to have the system working in its basic form before automating, otherwise, you are likely to automate your problems.

The demand pull system has proven itself far more efficient than the traditional push system. If the advantages of just-in-time are wanted, there is no alternative but to use a pull system, and kanban, in one form or another, is what is needed.

Visibility/Visual Control

One of JIT's great strengths is that it's a visual system. It can be difficult to keep track of what is going on in a traditional factory, with people hustling to and fro storing excess WIP, stored WIP being brought back to the floor for the next stage of processing, caches of buffer WIP all over the place, and the many crisscrossing production routes. The JIT factory is set up in such a way that confusion is removed from the system. In a JIT factory, it is easy to tell whether a line is working normally or having a problem. A quick visual scan reveals the presence of bottlenecks or excess capacity. In addition to the obvious signals, such as an idle workstation, JIT encourages the use of information boards to keep all the workers informed of status, problems, quality, and so on.

Each product work cell or team should have one or more boards, perhaps on easels, on which they post information. For example, if the schedule anticipates the production of 300 subassemblies for the day, the workers will check off the appropriate number each time a succeeding process pulls subassemblies from its output. This keeps the team apprised of how it is doing and presents the information to managers who only have to glance at the chart to gauge the work-cell activity and its kanbans to develop a very clear picture of how well the line is doing. Another board charts statistical process control data as the samples are taken in the work cell. Anyone can spot developing trends or confirm the well-being of the process with a quick look at the charts. Every time a problem beyond the control of the work cell, or an issue with which the work cell needs help comes up, it is jotted down on a board. It stays there until resolved. If it repeats before it is resolved, annotations are made in the form of four marks and a slash for a count of five (see Figure 14–13). This keeps the concerns of the work cell in front of the managers and engineers who have the responsibility for resolution. The mark tally also establishes a priority for resolution. The longest mark "bar" gets the highest priority. Maintenance schedules for tools and machines are also posted in plain view, usually right at or on the machine, and normal maintenance activity, such as lubrication, cleaning, and cutter replacement are assigned to the work cell.

Consider what happens when these charts are used. Information is immediately available to the work cell. The team is empowered to perform maintenance and solve all problems for which it has the capability. With the information presented to the team in real time, the team solves the problems at once and performs maintenance at appropriate times. This approach minimizes waste, keeps the machines in top shape, and produces a flow of ideas for improvement. The shop floor control loop is as tight as it can get. The operator detects and posts the information. The operator reacts to the information to solve problems or take action.

If a problem is beyond the work-cell team's capability, all the people who can bring skills or authority to bear are immediately brought in, presented with the data, and the problem gets solved—quickly. The control loop goes from information to action in one or two steps. In the traditional factory, the operator may not even be aware of a quality problem. It is usually detected by a quality assurance inspector hours or days after the defect was created. The inspector writes it up. The form goes to the management information system (MIS) department, where, after a period of time, the data are entered into the computer. Days or weeks later the computer prints a summary report including analysis of quality defects. The report is sent to management through the company mail or via an intranet system. The report may rest in an in-basket queue for a length of time before being examined. Managers in traditional plants are kept so busy with meetings and fire-fighting they hardly have time to read their mail, but eventually they will get around to looking at the report. They will see that the line is (or *was*) having a quality problem and pass the report to the floor supervisor for action. The floor supervisor will attempt to see whether the problem still exists. If it does not, case closed. It happened weeks ago, and the operator, who up until now was unaware of the defect(s), can't remember anything that would confirm the problem, let alone suggest a

Figure 14–13
Work-Cell Problem Status Board

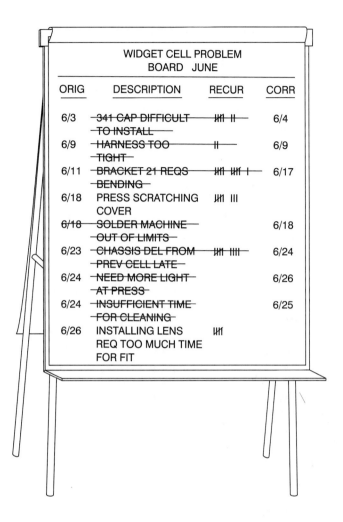

WIDGET CELL PROBLEM
BOARD JUNE

ORIG	DESCRIPTION	RECUR	CORR
6/3	~~341 CAP DIFFICULT TO INSTALL~~	~~卌 II~~	6/4
6/9	~~HARNESS TOO TIGHT~~	~~II~~	6/9
6/11	~~BRACKET 21 REQS BENDING~~	~~卌 卌 I~~	6/17
6/18	PRESS SCRATCHING COVER	卌 III	
~~6/18~~	~~SOLDER MACHINE OUT OF LIMITS~~		6/18
6/23	~~CHASSIS DEL FROM PREV CELL LATE~~	~~卌 卌I~~	6/24
6/24	~~NEED MORE LIGHT AT PRESS~~		6/26
6/24	~~INSUFFICIENT TIME FOR CLEANING~~		6/25
6/26	INSTALLING LENS REQ TOO MUCH TIME FOR FIT	卌	

root cause. If the floor supervisor is lucky, the problem may still be there, and the cause may be found. But in the meantime, weeks of production may have been defective.

In this control loop, at least six functions are involved before the loop is closed. That is bad enough, but when the time delay factor is added, any chance of finding root causes of problems that come and go is eliminated. Process improvement is much more difficult in this kind of traditional production system. Having had personal experience with both, the authors can attest that the most expensive, most sophisticated computer-based defect analysis system, such as might be employed in the above example, is infinitely inferior to the simple one- or two-step, person-to-person, no-computers-involved control loop of JIT when it comes to presenting useful information on a timely basis for the purpose of problem solving and process improvement.

Before our plants changed over to JIT, a mainframe-based defect analysis system was used. The Navy designated it as a best practice in the industry. Other companies came to see it, and many of them used it as a model for their own new systems. It could analyze data and present it in many different forms. But it had one flaw: time delay. From the time a process produced a defective part until the loop was closed with the operator of the process, several days (at best) had passed. We are not suggesting that the system was unable to make improvements, because it did. But the real revelation came with implementation of JIT and finding what could happen right inside the work cell when workers had the information they needed while it was fresh and vital, and they were empowered to do something with it. Immediately, defects dropped dramatically, and they continued to drop as continuous improvement was established. Before

JIT, these plants were never able to achieve results remotely comparable with their megadollar computer-based system.

Every JIT line develops its own versions of information display charts. But whatever the variation, everyone has valuable, useful information available at all times. That kind of information is extremely difficult to find in a traditional line and most often comes to light in the periodic (weekly or monthly) computer analysis reports. By then, the trail to the root cause may have become obliterated by the passage of time, other problems, or events. In the JIT factory, real-time visibility lets people know of the problem right then and there, while the cause is obvious. Coupled with the JIT philosophy that says that the problem must be solved before going any further, this visibility becomes a driver for elimination of problems, and for process improvement.

Elimination of Bottlenecks

Richard Schonberger makes the interesting point that only the bottlenecks in a traditional factory forward work to the next process just-in-time.[14] He explains that in a conventional manufacturing plant, the bottleneck process is one that goes as fast as it can all the time, barely keeping up with demand. If it breaks down, there is real trouble. To keep it running, and to attempt to find ways to increase its output, the bottleneck receives attention out of proportion to the rest of the plant, monopolizing the efforts of engineering and management.

In a JIT plant, all processes are potential bottlenecks in the sense just discussed, because there is little excess capacity, and there are no buffer stocks to fall back on when a process or machine shuts down. The upside of this is that all processes are constantly under scrutiny—none are ignored. As Schonberger also points out, the fact that all the processes must be watched carefully makes it imperative that the process operators play a major role in the care and monitoring and improving of the processes. There cannot be enough engineers to go around when every process is a potential bottleneck.

For this discussion, though, the bottleneck is put into a slightly different frame of reference. We are talking primarily about the setup stage of a JIT operation when trying to establish a balanced, rational flow through the production system. In this early stage, it is not uncommon to have some real functional bottlenecks. For example, if the new JIT line is being established to produce as many as 1,000 parts per day, but the manual assembly process can turn out only 800, there is a bottleneck. One way or the other, the process must be brought up to 1,000 or more. If the process employs two people using hand tools, then the answer is simple: add a third person and the appropriate tools. Then the capacity for that process should be 1,200 per day. The extra capacity will have to be accepted until the process can be improved to bring the daily single-operator output up to 500 each, making it possible to go back to two operators.

Perhaps a machine can produce only 75% of the projected demand. Here the options are a little different. This may be a very expensive machine, too expensive to replicate. Is it possible to put that machine to work somewhere else and put two, lower-capacity, less expensive machines in the line, or maybe a single new, higher-capacity machine? Can the old machine be modified to increase its output? If setup time is a part of the machine's normal day, then there is a potential for improvement. Another possibility may be adding a second smaller machine to augment the existing machine's capacity, although two different machines on the same line making the same part/product is not a desirable solution.

Another kind of bottleneck can exist when a single physical process is shared by two or more JIT lines. It is preferable to make each JIT product line independent and self-sufficient, but this is not always possible. An example might be a single-wave solder machine servicing two or more JIT lines. Because of the cost, size, and maintenance requirements of such a machine, it may not be feasible to put one in each JIT product line. Rather, all the JIT lines take their PC boards to a single-wave solder service cell for soldering. The JIT lines operate independently of each other. Therefore, it is difficult to predict when conflicts might develop. If they all need servicing at the same time, there is a bottleneck. If soldering delays cannot be accommodated, then one or more of the lines must have its own soldering capability.

Technology can often provide solutions to such problems. For example, 20 high-quality drag soldering machines could be purchased for the price of one wave soldering machine. Production rates of drag solder machines are much lower than those of wave machines, but in many applications they are ideal for placement right in the JIT line, dedicated to the line's product, and controlled by the line. Such solutions are often feasible with other types of machines too.

Whether your bottlenecks appear during the setup phase or during production, the best approach is to assign a cross-functional team to solve the problem. The team should have representation from engineering, manufacturing, finance, and any other relevant function areas. Its job is to list all possibilities for eliminating the bottleneck. This can be done by brainstorming, setting aside those ideas that don't make sense, and finding the most satisfactory solution in terms of quality, expense, efficiency, and timing.

Frequently, the solution to a bottleneck results in some degree of excess capacity in the process, as occurred earlier when the third operator was added. This is not always bad. Although JIT always works to achieve more and more efficiency and, taken to the extreme, would have just exactly enough capacity to produce the demanded level and no more—in a practical sense, some excess capacity is desirable. If a line is running at top speed every day, the operators will have no time for problem solving or improvement activities. Some time should be set aside each week for those two items, as well as for maintenance and housekeeping. For most applications, 10%–15% excess capacity is acceptable.

Small Lot Sizes and Reduced Setup Times

For 70 years or more, industrial engineers have been taught that the larger the production lot size, the greater the benefit from economy of scale. If one wanted to hold down cost of production, bigger lot sizes were the answer. This was the conventional thinking until the JIT manufacturing bombshell landed on our shores from Japan in the early 1980s. Under the leadership of Toyota and Taiichi Ohno, Japanese manufacturers concluded that the ideal lot size is not the largest but the smallest. Is it possible that both the manufacturers and the universities could have been wrong all those years? (In fact, many Western manufacturers and universities are still bound to the big lot philosophy.) Our conclusion is that the big lot was appropriate as long as mass production systems were used, although they certainly had major problems even then. But once the Toyota Production System came into being, the big lot was not only out of step but impossible to justify.

It stands to reason that if a machine is used to produce different parts that are used in the subsequent processes of production, and that the time it takes to change the machine over from one part type to another is 6 to 8 hours, then once the machine is set up for a particular part type, one should make the most of it. It seems to make more sense to run the machine with the same setup for 4 days, setting up for the next part on the 5th day, than to run 1 day, spend the next on setup, and so forth. The 1-day runs result in about 50% utilization time for the machine, assuming a single shift for simplicity. The 4-day run yields about 80% utilization.

So what is the problem? If there are four different parts to make on the machine, simply make 20 days' supply in 4 days, then go to the next part. By the time production has used all the 20-day supply of the first part, the machine will have cycled back to make that part again. Perhaps a 30-day supply should be made, just in case the machine breaks down. Would a 40-day supply be better? Where does this stop? If we are willing to risk an occasional breakdown, the 20-day cycle is acceptable. A place to store a 20-day supply of not just one part but four parts will be needed. Then the capability to inventory, retrieve, and transport these parts will also be needed. That represents land, facilities, and labor that would not be otherwise needed. None of it adds value to the product, so it is pure waste. It is likely that these costs add up to more than the supposed inefficiency of running the machine with a 50% utilization factor, but these costs are more acceptable to accountants. Land, buildings, and people in motion are not as apparent as examples of waste as machines that are not making product. Traditional thinking says, "Because the machine is busy, people are busy, floor space is full, it can't be waste." But it is.

In addition, suppose that a production flaw is found in one of the parts, caused by the machine. Every part made in that lot is suspect. Samples will be tested, and maybe the whole lot will have to be scrapped. This could be 20 days' supply, representing significant cost. The line will be down until new parts can be made—a major disruption.

Suppose the engineering department corrects a design weakness in one of the parts. Is the entire inventory of parts already made scrapped, or do we use them up in production, knowing that they are not as good as the newly designed part? Either is a bad proposition.

Now assume that the 1-day 50% utilization cycle on the machine was employed. The greatest loss we could take would be 8 days' inventory for any of these cases. The 8-day supply can be stored easier than 20 days' supply. This would reduce the cost of warehousing, control, and transportation. Any design changes can be cut in 8 days. Everything seems positive except the 50% machine utilization.

Ideally, setup time might be reduced to 30 minutes, producing 1 day's supply of each part every day. Utilization will be 75% and need for any warehousing may be eliminated. This may seem to be out of reach, but manufacturers using JIT have done far better, often taking setups from many hours to a few minutes. For example, by 1973 Toyota had reduced the setup time for a 1,000-ton press from 4 hours to 3 minutes. Over a 5-year period, Yanmar Diesel reduced the setup time for a machining line from over 9 hours to just 9 minutes.[15] These are not isolated examples.

The general rule seems to be that organizing properly for the setup, making sure the tools and parts that will be needed are in place, and having the right people there at the appointed time, will yield an immediate 50% reduction. Then by analyzing the setup process step by step, a company can usually streamline the process to cut time by half again. Ultimately, the machine itself may be modified to make setup faster and less difficult (by eliminating the need for adjustment). In any mature JIT factory, it would be a rare setup that took more than a few minutes, whereas the same setups were previously measured in hours.

Small lot sizes result in improved product quality, production flexibility, and customer responsiveness. Shortened setup times make small lots possible. The previously supposed advantage of manufacturing in big lots completely disappears when setup times are brought down to the kinds of times being discussed here. Machine utilization can be high to satisfy accounting criteria, and lots can be small to prevent waste and to enable kanban pulling straight from the machine to the next process. Short setup times coupled with kanban have the advantage of flexibility of production. For example, Harley-Davidson used to run its motorcycle line in long production runs of the same model. If a dealer placed an order for a model that had just finished its run, it might have been several weeks before that model could be run again, allowing the order to be filled. Harley was one of the first North American companies to adopt the total quality methods—as a means of survival.

For many years now, Harley has been able to mix models on the production line. They no longer have to produce their product in big lots because they were able to reduce setup times all along their line. Now when an order comes in, it is placed in the queue without regard for the model. Customers get their new bikes far sooner and are, therefore, less tempted by other brands in the meantime. Customer orientation is one major benefit of short setup times and small lot sizes, along with manufacturing flexibility, higher quality, and lower costs.

Total Productive Maintenance and Housekeeping

It is difficult to comprehend, but many manufacturers spend vast amounts on capital equipment and then ignore the machines until they self-destruct. By contrast, one can find relatively ancient machines in total quality Japanese factories that look like new and run even better. This must become the norm in the United States if U.S. companies are going to compete with the rest of the world. Because a JIT production line operates very close to capacity in every process, no tolerance exists for machine failure. When the machine is supposed to be running, it had better be, or the whole line will suffer. The Japanese have virtually eliminated machine breakdowns by applying total productive maintenance (TPM) techniques to their machines. Machines are cleaned and lubricated frequently, most of that work being done by the operators who run the machines. More technical preventive maintenance routines are performed by experts at frequent intervals. The machines are continually upgraded and modified for closer tolerances, faster setup, and fewer adjustments. Not only do the machines last longer, but during their entire life span they perform as well or better than when new.

The difficulty with TPM is finding the time in which to perform the maintenance, especially in factories in which three shifts are the norm during times of prosperity. The third shift is rare in Japan and Europe, so companies there do not share this problem. Regardless of the workday schedule, it is imperative that maintenance time be provided. The operator-performed maintenance is done during the normal shift (one reason to have a bit more than just enough capacity—a half-hour to an hour a day of excess capacity should more than cover the operator maintenance needs).

An added benefit of turning some of the maintenance responsibility over to the operators is that the operators develop a sense of ownership for the machines they use and care for. They pay keen attention to the looks, sounds, vibrations, and smells of the machines to spot problems before they develop. For the first time, the operators are in a position to call for maintenance before breakdown occurs. TPM is a must for JIT production systems.

Housekeeping is another area that is different under JIT. It is not unusual for the operators themselves to take on the responsibilities formerly associated with janitors. In the better JIT plants, one will see planned downtime being taken with cleaning chores—everything spotless, everything in its place. It follows that better performance will result from a clean, tidy, and well-organized work area than from one that is dirty and cluttered with tools scattered all over. People like a clean, bright, rational place in which to work. Again, time will have to be made available for this activity.

Process Capability, Statistical Process Control, and Continuous Improvement

Process capability, statistical process control, and continuous improvement have already been discussed in detail in this book, but it is important to understand the dependence on them by just-in-time. Is JIT a necessary prerequisite for process capability study and improvement, or for SPC, or continuous improvement? The answer is no. At least one of the three is being done in the majority of traditional production plants. Still, there is a connection. The philosophy and discipline of just-in-time virtually demand that they be used in any JIT environment. While a traditional manufacturing operation *may* employ one or more of the three, the JIT manufacturing operation *must,* and it must be all three. The reason may be obvious to you by now. The JIT plant is fragile. Everything must work when it is supposed to, and it must work close to perfection. There are no warehouses of buffer stock to come to the aid of a broken-down process. There is never much excess capacity to help out in tight spots. All the processes with their machines and people must operate in top form all the time.

This is where process capability, SPC, and continuous improvement come in. Even before the JIT line can be certified for full production, the line has to be balanced or rationalized, and a flow has to be established. Unless it is known what the processes are capable of doing in terms of quality and quantity, it will be difficult to achieve the even flow that is a necessary prerequisite of a kanban system. Without that, there is no JIT. In the traditional factory, it is not so much a problem: there is normally gross over-capacity, so parts are stored for the day things go wrong, and the bad parts are sorted out because there will still be good ones to use. In JIT, no extra parts can be made, and they all have to be good. Workers must have a handle on the processes.

Because one cannot afford (from the time or cost standpoints) to make defective parts, the processes must be in control at all times. The only way to assure this is through statistical process control. This is not as necessary in a traditional plant, but it is absolutely essential in JIT. Perfection is difficult to achieve in any circumstances, so it follows that in a complex manufacturing situation, perfection is next to impossible. This is certainly true. We never quite get to the point where all the parts are perfect, but with solid, stable, in-control processes forming the basis of a relentless continuous improvement program, we can come very, very close. (Some of the very best American plants target and achieve 6σ, three defects per million.) The best that can be achieved is the minimum that is acceptable for a JIT factory. In the process of continuous improvement, ways are found to do things better, faster, cheaper, and with constantly improving quality. The process never ends, and the diminishing return syndrome doesn't apply.

Suppliers

In the area of suppliers, JIT has different priorities from the traditional production system. The most obvious difference is the need for frequent, small lot deliveries of parts, supplies, and materials, rather than the traditional infrequent, huge volume deliveries. We are finding more and more JIT plants in which the suppliers deliver materials directly to the production cells, usually referred to as *point-of-use*. Several systems have been developed to cue the supplier that it is time to replenish materials. One is the *dual bin kanban* system. Two parts bins are used. Bin capacity may range from a few hours to a couple of weeks supply, depending on value, size, usage rate, and intended frequency of replenishment. When the cell has withdrawn all the parts from one bin, the empty bin itself is the signal that it is time to replenish. The supplier routinely checks the bins on the factory floor, and whenever he finds a bin empty, it is refilled with the exact number and kind of part designated on the bin label, usually in bar code. The supplier's bin checking must be scheduled frequently enough to assure that the second bin is never exhausted before the first is replenished. In a variation on the dual bin kanban scheme the cell's operators signal the supplier that a bin is empty, either by bar code transmission or automated electronic purchase order that is triggered by wanding the empty bin's bar code.

Clearly, for this kind of point-of-use materials delivery system to work, the supplier must by 100% reliable, the materials delivered must be of consistently high quality, and both the supplier and the manufacturing organization must be partners for the long haul. Consequently, choosing the suppliers for a JIT factory is a much more demanding job than it is for a traditional plant. Traditional factories are not so concerned with the delivery being on the dock at the precise date on the purchase order. It was going to be stored for a while anyway. And before that lot was used up, there would be another shipment in the warehouse. Neither do traditional factories concern themselves as much with quality from suppliers. The bad parts could always be sorted out, leaving enough good material to keep the line moving. The primary interest was price. Low price got the order. It quickly becomes apparent that this style of purchasing is incompatible with JIT.

The JIT plant must have its materials on the dock exactly on the day specified—in many cases at the hour and minute specified—or production may grind to a halt. Every part delivered must be a good part—there is no inventory cache from which to scrounge more parts to keep things moving. This means that the suppliers' quality must be consistently at or above specified requirements. Delivery and quality performance requirements of JIT effectively rule out buying for price. There is an oft-used phrase in JIT and TQM purchasing: "cost versus price." It suggests a holistic approach to the analysis of purchasing on the basis of total cost and value, not simply vendor price. How reliable is a particular vendor in terms of JIT deliveries? What kind of quality can be expected from the vendor? Does the vendor use JIT, SPC, and continuous improvement? Are its processes stable and in control? A supplier that gives positive responses in these and other areas may not be the lowest price contender but may well be the lowest cost. Value is what the JIT purchasing manager must look for, not lowest price on a bid sheet, because in JIT that turns out not to be the whole story.

When a JIT factory finds a supplier that delivers excellent materials on time, every time, there is every reason to want to continue to do business with it. More and more companies are turning to supplier partnerships to cement these relationships. What this means is that the two companies agree to work together, not only as supplier and customer but as unstructured partners. The JIT manufacturer may, for example, provide training and technical assistance to the supplier to get it started in total quality, JIT, SPC, and other processes. The JIT firm may certify the supplier's quality system to the extent that incoming inspections are eliminated, relying on the partner supplier to provide acceptable quality in all its deliveries. The supplier may be called on to assist in the design phase of a new product, bringing its unique expertise into the design team. Such relationships usually carry a multiyear agreement, so the supplier can count on the business as long as its performance remains high. There may be preferential bidding treatment, say, 10% advantage over non-partnership rivals. Effectively what happens is that the JIT manufacturer extends its factory right back into the supplier's. They operate to each other's requirements, and both are locked to each other. Results of this kind of arrangement have been excellent.

This kind of relationship is a far cry from the early ill-conceived attempts of some manufacturers to get into JIT before developing a full understanding of the concept. In those days, some companies would determine that by using JIT delivery of parts and materials, money could be saved. That part had some merit, but the execution was flawed. The companies simply told their suppliers to deliver a week's supply of materials once a week, rather than their customary 60 days' supply every 2 months. The suppliers' reaction is easy to imagine. They were being told, in effect, to store the materials in their own warehouses (which they didn't have) and to trickle the deliveries from the warehouses in small quantities weekly. This was simply a case of moving the storage facility from the manufacturer's plant to the suppliers'. A GM or a Ford has the power to do that to a supplier, but the suppliers, being smaller and with less influence, couldn't force the same back to their own suppliers, so they got caught in an intolerable situation. Only when the suppliers revolted and cried long and loud that this was not JIT—"and by the way, if you want me to store your goods for you, you're going to pay the tab anyway"—only then did the would-be JIT manufacturers see the error of their ways.

The new approach is working well because both parties benefit enormously. If a company wants JIT, then it must have the best possible suppliers, and both must want to work together for the long haul.

AUTOMATION AND JIT

Automation has not been discussed a great deal in this book. We have stuck to the fundamentals. One should not read into this, however, that JIT and automation are mutually exclusive. Rather, it is more meaningful to discuss the processes that use humans and manual machines, than the same processes powered by robots. If the fundamentals where humans apply are understood, the same fundamentals will be useful in an automated plant. All the same rules apply. We are not anti-automation.

We are, however, anti–"automation for the sake of automation." Many companies have made the costly mistake of thinking that automation will solve manufacturing problems. During the 1980s, manufacturers in the United States invested billions of dollars in automation. Cadillac built the most highly automated auto assembly plant in North America and probably the world. It turned into a nightmare of high-tech problems that took years to sort through. The plant that was to produce six cars per hour, after a year of operation, could do only half that and the quality of manufacture was questionable. Two years later, Toyota opened a new plant in Kentucky. Visitors to that plant, expecting to see a high-tech automated production line, were disappointed to find very little in the way of robotics.[16] The difference in the philosophies of the two companies becomes obvious. Executive managers at GM believed that by spending enough money, they could buy their way out of the trouble they were in. Toyota knew what it was capable of doing in one of its other low-tech plants that was operating successfully in Japan and simply cloned it down to the last detail in Kentucky. No razzle-dazzle, just good common sense.

Automation may be advantageous in many applications, but if you have not solved the problems in the human-operated versions of those same applications, you are not ready to automate them effectively. If you try, you will automate your problems and will find the robots far less adept at working around them than the humans they replaced.

It is frequently found that the need for automation is decreased or eliminated by converting to JIT. We certainly found that to be the case in two electronics plants. We were well into a program to build a factory of the future. The building was ready, much of the automation was on hand, and the rest—several million dollars worth—was on order when we started the conversion to JIT. Within months it had become obvious to everyone, including the designers of the new factory, that we were getting more out of JIT for almost no investment than could be projected for the new automated plant. The outstanding orders were canceled, penalties paid, and we walked away from the whole idea. We had learned in those few weeks of exposure to JIT that world-class manufacturing equates to JIT in a total quality environment, not to a factory full of robots and automatic guided vehicles. JIT and automation are compatible, but one should look long and hard at the need, and the company's readiness for it, before automating processes.

ENDNOTES

1. Taiichi Ohno, *Just-in-Time for Today and Tomorrow* (Cambridge, MA: Productivity Press, 1990), 2.
2. Ohno, 31.
3. Ohno, 28–29.
4. Ohno, 9.
5. Ohno, 75.
6. James Womack, Daniel T. Jones, and Daniel Roos, *The Machine That Changed the World* (New York: HarperCollins, 1990), 62.
7. David Lu, *Kanban—Just-in-Time at Toyota* (Cambridge, MA: Productivity Press and Japanese Management Association, 1986), vi.

8. Materials and general and accounting expenses are held constant for this example to keep it simple, although both could be expected to decrease under JIT.

9. C. E. Sorensen, *My Forty Years with Ford* (New York: Norton, 1956), 174.

10. Womack, 83.

11. Lu, 73.

12. M. Scott Myers, *Every Employee Is a Manager* (San Diego: Pfeiffer, 1991), 72.

13. Lu, 87–92.

14. Richard J. Schonberger, *World Class Manufacturing* (New York: Free Press, 1986), 67.

15. Kiyoshi Suzaki, *The New Manufacturing Challenge: Techniques for Continuous Improvement* (New York: Free Press, 1987), 43.

16. Maryann Keller, *Rude Awakening: The Rise, Fall, and Struggle for Recovery of General Motors* (New York: Morrow, 1989), 206–209.

Index